**Hobby Publications**

Birds of
Marshside

Barry McCarthy

Text by
Barry McCarthy

Illustrations by
Neill Hunt, Sebastian Hunt, John Dempsey, Barry McCarthy and Bill Morton

Cover Illustration
Angela Baker 'Seasons at Marshside'. Photography by Mike McKavett.

Designed and typeset by
Twentyten Media Ltd

First published by
Hobby Publications
c/o Twentyten Media Ltd
Red Lion Building
1 Liverpool Road North
Maghull
Liverpool L31 2HB
United Kingdom

Tel: 0151 526 7788
Fax: 0151 526 7799

© Hobby Publications 2001

Printed with the kind support of
Ashley Printers Limited, Liverpool
Tel: 0151 236 4415

British Library Cataloguing-in-Publication Data.
A British Library CIP Record is available.

ISBN 1-872839 07 X

All rights reserved.
No part of this publication may be reproduced, stored in a retrieval system, or transmitted, in any form or by any means, electronic, mechanical, recording or otherwise without prior permission of the publishers.

Contents

6 **Preface**

7 **History**

8 **Introduction**

11 **RSPB Reserve Rankings**

12 **Systematic List of Species**
   Recorded to 31st December 1999

113 **References**

Preface

In this book I have attempted to present a detailed account of the birds of Marshside and Crossens, concentrating upon the status of each species at the end of the 1990s, but referring where appropriate to the past, mainly from the 1970s onwards. It was originally hoped to produce a full natural history of this rich area, but that project will have to await a more widely knowledgeable author, or better still a team of specialists.

In 1995, Tony Baker, warden of the newly-created RSPB reserve, suggested that an account of the birds of Marshside and Crossens would provide both helpful information for birders and also serve as a valuable baseline with which to compare the inevitable changes of the coming years. The project took much longer than I had anticipated; an enormous amount of uncollated data was unearthed in field notes stretching back over several decades, not to mention the mass of records reported in the county and other bird reports over the years.

It was finally decided to bring the story up to date, as of the end of the 20th century on 31st December 1999. I have resisted the temptation to include new information for 2000, including our first record of Cattle Egret and the first confirmed breeding of Teal, in the sure and certain knowledge that the project would never end unless a definite cut-off point was applied.

I wish to acknowledge the contributions of the following birders who have provided information, advice, encouragement, companionship in the field, or in many instances all four:

Mike Ainscough, Pete Allen, Tony Baker, John Bannon, Pete Carty, John Dempsey, Mark Garner, Ross Hughes, Neill Hunt, Simon Jackson, Maurice Jones, Mike McKavett, Steve Riley, the late Malcolm Rimmer, Harry Shorrock, Mike Stocker, Paul Thomason, Steve White and Brian Woolley.

Particular thanks are due to the publisher John Bannon, who drove the project on to completion when my own energies were flagging; to Tony Baker who suggested the idea in the first place and to Maurice Jones, the Lancs county recorder who made available old and rare issues of the Lancashire Bird Report; also to Neill and Sebastian Hunt, John Dempsey and Billy Morton whose vignettes along with some of my own, enliven the text throughout.

Any errors or omissions in the the species accounts are mine and mine alone; I dedicate this publication to all Marshside birders, past, present and future.

Barry McCarthy
December 2000

Marshside ...a brief history

What is now the RSPB Reserve at Marshside was, until only recently, very much part of the often turbulent Irish Sea and it was not until the completion of the new coastal road in the early 1970s that the tidal saltmarsh was finally cut off from the sea.

The coastline north of Southport as far as Crossens was formerly bordered by extensive sandhills, the old Norse word for these dunes being 'meols' and Churchtown, the oldest established settlement in the area is listed as Meols in the Domesday book. Remnants of these high dunes can still be seen today on the adjacent Hesketh Golf Club, as the first nine holes are laid out through the original sandhills.

In former days high spring tides reached the very doorsteps of Churchtown, nowadays some two kilometres inland and Marshside itself was but a few scattered fishermens' cottages between the' meols' and the sea. Slowly but inexorably the 'dry land' extended seawards, assisted by the construction of sea-cops and embankments, until the entire area of the current reserve was impounded between the19th century sea defences and the coast road.

Over the years, the sea has made several serious attempts to reclaim its' former territory, as in the 'great inundation' of 1720 when tens of thousands of acres were flooded to a depth of six feet or more. Even nowadays the present coast road is seriously undermined by the combined effects of violent westerly gales and high tides, although the same storms do bring in rarer seabirds such as Leach's Petrel.

In 1869, seven Marshside shrimpers were drowned when a sudden sea fog shrouded their horse-drawn carts and completely disorientated they were overcome by the incoming tide. As a result a fog bell tower was built which still stands today on Marshside Road, just inland of the reserve boundary: a visible link to those former seafaring days when Marshside was full of boatyards, sailmakers and net-repairers as befits an active fishing community on the shores of the Celtic sea.

John Bannon
December 2000

A Fragment of a Wilderness

Marshside in the context of the Ribble Estuary and the Martin Mere mosslands

This book describes a site of modest size yet one which encompasses within its bounds the essence of the Ribble Wetlands. Once a natural succession of habitats from mudflats and saltmarsh to grassland, swamp and mere spread for up to twenty miles inland. Now the freshwater habitats have mostly been drained and what remains is much modified by Man, like the RSPB reserve at Marshside itself.

In these small fragments, at Marshside and at Martin Mere, the naturalist catches a glimpse of the wetland riches that must have awed our ancestors. The swamps that once stretched from the Fylde to the Mersey and across the Douglas may have been tamed and drained but we can still rejoice in a wetland wildlife that is diverse and resilient. Even now, if we were to give Nature just a little encouragement, we might yet recreate a part of that wonderful, watery vision.

The RSPB at Marshside

The story of RSPB involvement at Marshside like so many green issues in the 1960's and 70's began with conflict, and it was the conservationists who began it as the underdogs. Southport needed to develop and, trapped as it is between the retreating sea and the commercially valuable agricultural land of the Mosses, it naturally looked first to the drying `waste-ground' of the marshes for space to build. Many politicians at that time saw Marshside as a natural site for development. The conservationists had other priorities and were determined that the bird-life of the marshes should be protected at all costs.

Twenty years later, the climate has changed in more ways than one, and many of the new generation of politicians are themselves committed conservationists . Happily, the current Local Authority , Sefton Council , has an international reputation for its coastal management expertise and has welcomed the RSPB as the appropriate body to manage its premier coastal wetland for birds

Marshside's Importance for Birds

The RSPB reserve is just a small part of the Ribble Estuary Special Protection Area (SPA). This is a European designation that recognises the Ribble's International importance for waterfowl. Indeed, it is the second most important wetland in the UK for wintering waterfowl and is only regularly surpassed in numbers by The Wash.

Marshside constitutes only a small fraction of the whole estuary but its importance to the birds is out of all proportion to its size. The reserve is just 280 acres (110 ha.) in extent but the numbers of birds present in the winter qualify it as internationally important. It regularly holds a total of over 40,000 birds which includes over 1% of the European population of black-tailed godwit and pink-footed goose and over 1% of the National population of wigeon, teal, shoveler, golden plover, ruff, and herring gull.

Marshside's breeding birds are also receiving increasing interest as populations of breeding waders decline remorselessly across the UK. The lapwing colony is one of the most significant of any RSPB reserve. Together with the healthy redshank population the total count of breeding waders gives it the highest density of any RSPB reserve at approximately one pair per hectare.

The key to Marshside's importance is its habitat: coastal grassland. Free from tidal inundation yet still close to the estuary the reserve's vital ingredients of fresh water and summer-grazed grassland are attractive to a wide range of species. Many, like the golden plover, feed both on the grassland of the reserve and out on the estuary's mudflats whilst others, like the shoveler, may spend all their time on the reserve's open water. A few species, like the bar-tailed godwit, only venture onto the reserve during very high spring tides when they have no other option. Nevertheless, the reserve is then a vital part of their ecosystem; they literally have nowhere else to go and, during a cold winter snap, finding a safe place to rest for an hour or two may mean the difference between life and death. This coastal grassland habitat is increasingly scarce in the UK and Marshside is one of only two remnants left on the Ribble. Thousands of acres have gone under the plough or suffered an even worse fate as sites for our domestic waste.

Planning for the Future

The first RSPB Management Plan was written in 1995 and guided the first five years of the reserve's development. It began by examining the wealth of bird life and identified those species for which the new reserve was of International importance (pink-footed goose and black-tailed godwit) or National importance (teal, shoveler, golden plover, ringed plover, ruff and herring gull). Marshside's breeding birds are currently only of regional importance but consist almost entirely of those species for which there is most conservation concern as modern agriculture takes its toll on their traditional farmland haunts.

So what can be done to improve the lot of the most important species at Marshside? For the wintering birds, freedom from disturbance, a suitably short sward and plenty of water are the main requirements. These are relatively easily met here on the wetter side of Britain. The breeding birds have similar needs but maintaining plenty of water through the breeding season on a sandy site like Marshside has proved more difficult.

This has been the main focus of RSPB management to date and considerable progress has been made in keeping the waders and wildfowl productive through the driest of breeding seasons. Unfortunately, the reserve cannot function in isolation and some species may continue to decline as the greater part of their range continues to deteriorate, at least in the short-term. The decline in linnet and skylark numbers, for example, is largely due to the `efficiency' of modern farming methods which have deprived the birds of the seed from winter stubbles and the insect life of summer crops.

The second five year Management Plan is about to come into force and will build upon the successes of the first. As for those external factors upon which our small birds depend, we must hope that the politicians will also continue to build on the recent successes of environmentally-friendly farming initiatives like MAFF's `Countryside Stewardship' scheme. This scheme currently supports the traditional farming methods used at Marshside and could be used to great benefit throughout the Ribble Wetlands and beyond.

Away from the RSPB reserve there are positive signs in all directions. The Hesketh Golf Club have been involved in work to improve the fortunes of both the natterjack toad and the sand lizard and may be able to help us improve our bird populations in the future too. The saltmarshes and sand-flats bordering the RSPB reserve are owned by Sefton Council and plans are afoot to designate the area a Local Nature Reserve with the management emphasis being the welfare of the important bird populations. The pink-footed geese seem to have already registered their approval by using the saltmarsh around the sand-works with increasing regularity. To the north, the bulk of the estuary is secure in the hands of English Nature, so the future of this part of the world does indeed look relatively rosy.

The New Millennium at Marshside

With the next five year Management Plan about to take effect, `The Birds of Marshside" is a very timely publication, summarising as it does our current knowledge of Marshside's birds. But what can we expect in the future?

Too many things are changing too fast to be very confident in predicting the future fortunes of our birds. Climate change and the Government's agricultural, energy and transport policies are obviously going to be very important and probably much more important than anything that we can do on the Nature Reserve. We do now have the benefit of a great deal of knowledge of the sort you will find in these pages, and an increasing number of dedicated and increasingly expert birders. With the influence of the RSPB and the other conservation bodies at such a high there must be cause for some optimism that our birds will continue to thrive into the future. Forgive me if I take this opportunity to remind everyone that the grounds for that optimism depends entirely upon you all voting with your wallets and making sure that you are supporting the RSPB!

With a copy of 'The Birds of Marshside' on our shelves we can now take to the field helping to chart the progress of our birds in the future. No doubt there will be losses but there is much cause for optimism too. A new breeding or wintering species, a first for Marshside, a first for Lancashire. Who knows? - a first for Britain is always possible! Happy birding!

Tony Baker
Site Manager
Marshside RSPB Nature Reserve

RSPB RESERVES RANKING IN THEIR CONSERVATION IMPORTANCE FOR WINTERING AND PASSAGE WATERFOWL

RANK	RESERVE	INTN'L	GB & IRL	SCORE	SIZE (Hectares)
1	Elmley Marshes (Kent)	2	10	26	287
2	Ouse Washes (Cambs/Norfolk)	7	3	25	1046
3	**Marshside (Merseyside)**	**2**	**8**	**22**	**110**
4	Dee Estuary (Merseyside/Dee)	3	6	21	5415
5	Lough Foyle (Londonderry)	4	5	21	1335
6	Snettisham (Norfolk)	3	7	21	1630
7	Nene Washes (Cambs)	4	4	20	328
8	West Sedgemoor (Somerset)	4	2	16	576
9	Titchwell (Norfolk)	3	3	15	379
10	Langstone Harbour (Hants)	1	5	13	554

Scoring for Species
Internationally Important (Amber) 3
GB/All Ireland Important (Amber) 2
GB/All Ireland Important (Not Amber) 1

Adapted from Birds of Conservation Concern and other Key Species at RSPB Reserves
by J Cadbury and J Evans RSPB 1999

Systematic List of Species

Recorded at Marshside up to the end of 1999

RED-THROATED DIVER *Gavia stellata*

One on the high tide off the Sand Plant peninsula on 11 September 1992 and two together there on 8 October 1995 are the only records. It is likely that the Red-throated Diver is under-recorded on the estuary, given that the species is a regular passage migrant and winterer off both Formby Point and the Fylde coast.

BLACK-THROATED DIVER *Gavia arctica*

One on Crossens channel on 7 October 1967 is the only record.

LITTLE GREBE *Tachybaptus ruficollis*

Until 1989 this species was an infrequent visitor to the Sand Plant lagoons and adjacent creeks in autumn and winter. In April 1989 one or two birds spent several days on Polly's Creek; breeding was not suspected but there were a few sightings of individual birds during the following winter and in the spring of 1990. In 1991 breeding commenced with a single pair raising at least two young on M2. This was repeated in 1992, increasing during 1993-95 to two pairs. In 1994, 1995, 1998 and 1999 attempts by a third pair to breed on M1 failed, in 1995, at least, due to the rapid drying out of the marsh in late May. Severe drought conditions almost certainly led to failure of the single brood on M2 in 1996, but in 1997-99 at least one pair per year was again successful in raising offspring.

Since breeding has become established Little Grebes may be encountered at any time of year, usually in the Sand Plant lagoons-Polly's Creek area but also in late autumn and winter on the M1 floods, and occasionally even on the small ponds on Hesketh Golf Course.

GREAT CRESTED GREBE *Podiceps cristatus*

Occasional offshore on high tides in autumn, winter and spring. The species is recorded in most years and usually occurs in ones and twos; 11 off the Sand Plant peninsula on 27 December 1992 and nine on 11 December 1996 are the highest numbers recorded. One on the Sand Plant lagoons on 7 November 1998 was the first to be recorded inland of the Marine Drive for at least 20 years.

SLAVONIAN GREBE *Podiceps auritus*

The only documented record is of a single bird offshore on 21 October 1978.

BLACK-NECKED GREBE *Podiceps nigricollis*

One on the tide off Crossens on 1 April 1974 is the only record.

FULMAR *Fulmarus glacialis*

Fulmars are recorded very infrequently at Marshside, mainly in autumn. Two were on the Sand Plant lagoons after gales on 16 September 1990, an exhausted individual was picked up in the Corral area of M2 on 26 September 1995 and subsequently released, and another was on M2 after overnight gales on 17 September 1998. Other sightings have been offshore with two in a violent gale on Christmas Day 1999 the most recent.

LEACH'S PETREL *Oceanodroma leucorhoa*

This is a regular autumn migrant on the Lancashire coast, sometimes in considerable numbers during westerly gales, but only very rarely reported from Marshside. Eight on 7 October 1980 and singles on 7 September 1986, and on 3 January and 28 October 1998, all during or just after stormy weather, are the only recent records. The species may well be under-recorded, as the weather conducive to the appearance of Leach's Petrels is likely to lure birders away to established sea-watching sites such as Formby Point, Seaforth or Blackpool.

GANNET *Morus bassanus*

There are, surprisingly, only a few records of Gannets offshore at Marshside in summer and autumn, three on 22 August 1994, two on 26 July 1998 and four in late September 1999 being the most recent occurrences. Given the often very high numbers moving off both Blackpool and Formby Point between mid-June and late September, with counts of over 100 quite common, it is probable that Gannets are under-recorded at Marshside, passing well offshore beyond the attention of most birders.

CORMORANT *Phalacrocorax carbo*

Cormorants are regular visitors which may be met with at any time of year. One or two birds are often seen fishing the Sand Plant lagoons or Polly's Creek, where they appear to find enough Eels and flatfishes to hold their interest for a few days. Most sightings are in the period October to April and most are of immature birds. Cormorants are also quite frequently seen offshore, usually singly or in small groups flying up or down the Ribble at high tide, or resting on the sandbanks; numbers have grown rapidly since 1997, and 43 on 2 January 1999 is the highest total so far recorded.

BITTERN *Botaurus stellaris*

There is only one record, of a bird flying inland over Crossens Inner on the evening of 22 October 1997.

NIGHT HERON *Nycticorax nycticorax*

A second-year and a third-year frequented ditches on Hesketh Golf Course each evening from 13 to 16 April 1980, the older bird remaining until 22 April and roosting by day in conifers in nearby Hesketh Park. Another third-year bird was in the same area of the Golf Course on 14 and 15 May 1983.

LITTLE EGRET *Egretta garzetta*

There are eight Marshside records of this increasingly regular visitor to North-West England, involving at least 11 birds. During a substantial influx into the region at least three and probably five immatures were present on 9 August 1993, one remaining until 13 August and frequenting both salt and fresh marshes. A different immature individual was at the now silted-up pool beside the rubble road used by the sand trucks on 31 August 1993. One was on the saltmarsh by the Sand Plant peninsula briefly on 27 May 1996, and another spent several hours in the same area on 25 May 1997; there was a brief visit to the saltmarsh by a single bird on 4 October 1997. One was seen on the saltmarsh on several dates in late August and early September 1998, and one was on M2 on 11 May 1999. The most recent occurrence is of two birds that frequented the tideline and creeks off the peninsula on numerous occasions between 28 August and 30 September 1999.

GREY HERON *Ardea cinerea*

Grey Herons are regularly present in varying numbers on both fresh and saltmarshes. Up to five or six may be seen on almost any visit, and both adult and immature birds are usually present; 23 in January-February 1980 and 17 in July 1983 are the highest counts recorded in recent years. A melanistic individual was seen on several dates in July and August 1993.

The origins of our Grey Herons are not known; it is assumed that most are visitors from breeding sites in nearby mossland woods.

PURPLE HERON *Ardea purpurea*

An adult or near-adult flew north-west over Crossens on 9 May 1979.

WHITE STORK *Ciconia ciconia*

There are two records of single birds in the last 40 years, both in early spring: one passed over on 4 April 1961 and another flew north at midday on 31 March 1984. The latter bird was almost certainly the same individual seen at Pilling in the Fylde later in the day and relocated early on 1 April before it moved off again eastwards.

SPOONBILL *Platalea leucorodia*

An adult fed briefly in a ditch by the Sand Plant lagoons on 26 May 1991 before moving off south; also in 1991 two immatures were present intermittently on the shore and saltmarsh creeks from 6 to 12 October. One visited both fresh and saltmarshes on several dates between 27 May and 15 June 1996. 1999 saw a surge of records, with many sightings of a single bird on M2 in late April and again between June and mid-August; it is not clear whether more than one bird was involved in the series of records.

MUTE SWAN *Cygnus olor*

The Mute Swan is a frequent visitor in winter and spring to the M1 floods and the Sand Plant lagoons/M2 creeks. Numbers vary; four to six is typical, but up to 46 in December 1992 and 48 in April 1998 are the highest counts recorded. Numbers have tended to increase since the early 1990s in line with the growth in the population on nearby Southport Marine Lake from which many of our birds are known to commute.

The breeding history of this species has been erratic. A pair nested by Crossens channel intermittently during the 1970s but not apparently since. Up to two pairs have attempted to breed on M2 or by the Sand Plant lagoons each season since 1992 with limited success, and a single pair has nested successfully on Hesketh Golf Course each year since 1993. Most of these various efforts have been relatively unproductive in terms of numbers of offspring raised; one or two cygnets is the norm with seven the highest total recorded.

BEWICK'S SWAN *Cygnus columbianus bewickii*

Up to about 1970 this species was an early spring passage migrant in small but gradually-increasing numbers, seen mainly on Crossens saltmarsh (Greenhalgh, 1975). In the mid-1970s the status of Bewick's Swan changed dramatically as a substantial wintering flock became established at Crossens. Birds normally began to arrive in early December, built up to a peak in January, maintained fairly high numbers throughout February, and then departed quite quickly from early March onwards. Both the saltmarsh and Crossens Inner were frequented and up to the mid-1980s this flock was one of the premier winter birding spectacles in the region. Numbers increased steadily over the period: in 1978 peak counts were 100 in January, 106 in February, 122 in March and 126 in December; in 1984, 330 in January, 317 in February, 270 in March and 295 in December.

The end, when it came, was even more sudden than the beginning: totals in the next winter, 1985-86, were somewhat lower than before, peaking at 182 in December and 200 in early January; only five birds returned in late December 1986 and these departed early in January 1987. From then until December 1989 there were occasional one-off sightings of family groups during mid-winter but from 1990 to 1993 the only record was of a single flock of some 50 birds on 6 December 1992. Local birding opinion attributes the disappearance of the Crossens swan flock to the initiation of an artificial wildfowl-feeding programme at Martin Mere WWT reserve which commenced at about the same time; this certainly seems plausible, as numbers of wintering Bewick's Swans increased rapidly at Martin Mere in the mid-1980s and have remained high ever since.

Fortunately, however, the story of Bewick's Swans at Marshside may not yet be over: in January and February 1994 small parties began to visit the winter floods on M1, peaking at 37 on 15 January. This pattern has continued in all subsequent winters with brief visitations, mostly by small groups, between mid-December and mid-March. A flock of 150 on the M1 floods on 25 January 1998 was the largest count at Marshside for over a decade.

WHOOPER SWAN *Cygnus cygnus*

The recent history of the Whooper Swan at Marshside parallels that of the previous species but as Whoopers have never attained the numbers reached by Bewick's Swans the changes in their status over the years have seemed less dramatic. Greenhalgh (1975) describes Whooper Swans as rare on the Ribble estuary as a whole before about 1970 with a gradual increase to around eight birds each winter by 1974-75. At Marshside a single bird accompanied the Bewick's Swan flock in the 1976-77 season, but regular wintering began in 1978-79 when from 14 to 30 Whoopers were with the swan flock until late March with several remaining until mid-April, long after all of the smaller species had left. This set the pattern for the next few winters: a handful of Whoopers arriving with the Bewick's Swans in December and staying on until late March or early April; peak count was 15 in January 1981. Numbers increased substantially in January 1983 when 35 birds were on Crossens and up to 45 returned for the 1983-84 winter; it may not be mere coincidence that Bewick's Swan numbers also reached their peak during these two winters.

Coincidence or not, the subsequent decline in Whooper numbers was extremely rapid: 15 birds on 19 December 1984 were the last to be recorded at Marshside for eight years until 31 were seen on Crossens saltmarsh on 6 December 1992! Apart from a single bird on New Year's Day, 1993 was another blank year but on 16 January 1994 five were on the M1 floods and up to 130 spent over two weeks on Crossens saltmarsh in December. A few also visited the floods in February-April 1995, up to 23 were on the saltings on 24 December 1996 and 30 on 14 December 1998, so a modest recovery similar to that in the fortunes of the Bewick's Swan may be underway.

BEAN GOOSE *Anser fabalis*

Bean Geese are eagerly looked for among the Pink-footed Geese at Marshside each winter by visiting birders and found with a frequency that has long amazed the local regulars. In fact this species is a scarce and quite irregular winter visitor; there are very few records involving more than one bird at a time and several successive years may pass without a single sighting. When they do occur birds seem to move about apparently at random in the South-West Lancashire area and may spend only a few days or even only a few hours at Marshside along with their chosen Pinkfoot flock. Examination of the records over the past three decades reveals some interesting patterns.

First, good 'Bean Goose winters' seem to occur in clusters: apart from one shot at Marshside in December 1968 the late 1960s seems to have been a lean period, followed by a good series of records from 1972 to 1975 (Greenhalgh,1975). Except for one in January 1977 there were no records at Marshside from 1976 to 1980 although birds were recorded on the mosslands during some of those winters. Two were seen in February-March 1982 but none in 1983 or 1984; from 1985 to 1987 another run of good years produced records both in January-March and in December, with up to ten birds present in February-March 1985. After a relatively thin period in 1988-1992 there was an increase in records in 1993-94 involving up to three individuals in each year; one bird spent some five weeks, mainly on M1, in November-December 1994. After a lull in 1995 one was on M1 on 8-9 January 1996 and another on 23-24 November 1997.

A second discernible pattern concerns the dates of occurrences: up to about 1985 the great majority of records were in the period January-March but since then an increasing and now predominant number have been early in the winter, from late October onwards. This change may merely be a by-product of a similar recent tendency by the flocks of Pink-footed Geese, which the Bean Geese accompany, to feed at Marshside earlier in the winter than formerly and then largely to leave the site in December or early January. Most of our Bean Geese have historically been assigned to the northern race *fabalis*, nearly twice as many as to the eastern form *rossicus*.

PINK-FOOTED GOOSE *Anser brachyrhynchus*

Since the early years of the present century at least, the Pink-footed Goose has been the commonest goose wintering on the Ribble estuary and on the South-West Lancashire mosslands in general (Greenhalgh,1975). Numbers have increased steadily

from around 2,000 in the late 1930s to some 4,000 in the 1950s and to 14,000 by the mid-1970s. By the early 1990s over 30,000 birds were being recorded in systematic counts each winter (Cranswick et al, 1995). At Marshside and Crossens, in particular, counts of up to 10,000 Pinkfeet had been reached by December 1980. By mid-January 1986 a peak of 15,000 was recorded, all on the fresh marsh inland of the Marine Drive, and up to 22,500 were on Marshside-Crossens-Banks marshes combined on 10 November 1991. Although this last count has never been equalled an impressive 13,000-plus were concentrated on M2 and Crossens Inner in late November 1994.

Pink-footed Geese begin to return from their breeding grounds in central Iceland in September. In some years the first skeins are seen as early as the 10th of the month but more usually it is late September before any appreciable flocks are encountered. Numbers build up very rapidly through October and early November; since about 1987 the peak winter counts for Marshside-Crossens have nearly all been in November-early December, a significant change from earlier years when January and February normally saw the highest numbers. As recently as the early 1990s it was by no means unusual for up to 3,000 Pinkfeet to remain, usually on the saltmarsh, until mid-March; since about 1992, however, the flocks have departed ever earlier in the New Year and in 1994, 1995, 1998 and 1999 virtually all had left Marshside by early January.

Although the causes of this change in the behaviour of the flocks are no doubt multiple and complex, factors such as the recent run of mild winters and changes in availability of various food resources as winter progresses presumably play a role. It has been suggested that some at least of the

Lancashire Pinkfeet may have moved to East Anglia in late winter during recent years. In spite of the trend towards an early departure exceptional winters do occur; in 1996-97 up to 7,000 were to be seen on both salt and fresh marshes up to the end of January. Small groups of transients are often recorded well into May at Marshside, and in some seasons one or more birds, almost certainly injured, linger on through the summer. Occasionally, leucistic individuals and birds with orange legs turn up among the winter flocks to puzzle the novice.

WHITE-FRONTED GOOSE *Anser albifrons*

According to Oakes (1953) the nominate Russian race was a winter visitor in some numbers to Lancashire as far north as the Ribble during the early decades of the present century. By the 1940s, however, numbers had declined and by the mid-1960s no more than a handful were recorded among the Pinkfoot flocks in South-West Lancashire in most winters (Greenhalgh,1975). At Marshside Russian Whitefronts have long been scarce visitors although numbers can vary considerably from year to year. Very few were recorded during the 1970s but an average of three or four each winter in the early 1980s gave way to 20 in late January-February 1985; a flock of 36 in early December 1987 had increased to 95 by year's end and this party remained into February 1988.

During the next four winters *albifrons* Whitefronts were again scarce at Marshside with no records at all in 1989 and a peak count of eight together on 24 November 1990. In February 1993, however, a large cold-weather influx into Britain from the near Continent brought a flock of 23 to Marshside which remained well into March. A group of six birds with the goose flock from 20 November to 3 December signalled a return to more usual numbers in 1994. Up to seven were present in December 1995, and this nucleus increased to over 30 by early March 1996; since then numbers have been low, with no more than one or two recorded in a typical winter.

The Greenland subspecies flavirostris has generally occurred in much smaller numbers than its European relative. One or two are seen in most winters but quite a few years have passed without a sighting since the mid-1970s; six in late December 1987 and up to 22 in late November 1989 are the highest counts on record. As one would expect from the geographically very distinct origins of the two races there is no correlation between the occurrences of albifrons and of flavirostris among the goose flocks, although the situation at Marshside is confused by the fact that the same geese have on occasion been reported as *albifrons* by some observers and as *flavirostris* by others.

LESSER WHITE-FRONTED GOOSE *Anser erythropus*

A first-winter bird on the saltmarsh with Pink-footed Geese for several hours on the morning of 19 January 1985 is the only documented record.

GREYLAG GOOSE *Anser anser*

Small flocks of 'wild' Greylag Geese wintered on the Inner Ribble marshes particularly at Freckleton and Longton until the mid-1950s, but numbers had declined to mere ones and twos by 1965 (Greenhalgh, 1975). At Marshside, apart from a good season in 1978-79 when up to 28 were among the Pinkfeet in January-February, very small numbers of 'wild' Greylags accompanied the commoner species during several winters in the 1970s and early 1980s. Since then although 'wild' birds continued to be claimed at least until 1994 the presence of an increasing feral flock makes the validity of these later claims difficult to assess.

The first feral Greylag Geese were recorded at Marshside in 1985 when up to 12 birds were present around the Sand Plant lagoons and on M2 for lengthy periods in April, June, September and December. The immediate origin of our feral flock remains unknown but the species has increased on many waters in the area since the 1970s. Numbers at Marshside grew quite slowly in the early years, although first breeding took place in 1987 when a single pair reared 11 goslings by the Sand Plant lagoons. Peak counts rarely exceeded 30 birds until October 1991 when up to 76 suddenly appeared on M2 and Crossens Inner; most remained until December. Since then a flock of up to 50 Greylags has become a regular feature of birding at Marshside at almost any time of the year with peak counts of 90-140 birds usually in the period September-December. Breeding by a single pair occurs about every other year and productivity has been fairly low in recent years with four the maximum number of offspring reared at any attempt; two pairs nested successfully in 1999, however.

SNOW GOOSE *Anser caerulescens*

The Snow Goose has been recorded as a straggler among the South-West Lancashire goose flocks since at least the late 1950s and the debate over its status, wild American vagrant or escape from captivity, has been going on ever since, surfacing again whenever a new individual turns up at Marshside or on the mosslands. The issue seems essentially unresolvable with too many unquantified variables to permit more than an estimate of probabilities in each case; most of the local regulars are willing to accept Snow Geese seen among the Pinkfeet in winter as wild birds unless there are specific reasons to think otherwise.

At Marshside most records have been of the Lesser Snow Goose *A.c. caerulescens* which can occur in several gradations of a 'blue' phase as well as in the typical white form (Ogilvie,1978). There are a few scattered occurrences over the past four decades: a blue-phase bird on several dates from late October to early December 1958, a white morph in January 1969, another white bird in December 1970-January 1971 and perhaps the same individual briefly during the 1971-72 winter, a white-phase on 12 December 1989, and a blue goose on many dates between 17 October and 14 November 1992 and again from 3 to 9 December of that year. The Greater Snow Goose, *A.c. atlanticus*, which does not have a blue phase, has been represented at Marshside by three individuals: one which was present with the Pink-footed Geese intermittently from 10 November 1984 to 14 February 1985 and is assumed to be the bird that returned for brief visits during the 1985-86, 1986-87 and 1989-90 winters; a second individual present on several dates in December 1991, and another on several dates in November 1996. One or two white feral geese which regularly accompany the Greylag and Canada Goose flocks have produced many misidentifications of Snow Geese in recent winters.

CANADA GOOSE *Branta canadensis*

The Canada Goose probably shares with its cousin the Greylag the distinction of being Marshside's most unloved bird, among birders, at least. In fact, however, there is in addition to the recently-arrived feral flock a number of records of 'small race' Canada Geese which may refer to genuinely wild birds from arctic North America that have found their way to these shores in the company of Icelandic Pinkfeet. The most recent of these sightings are of single birds on 16-26 November and 14 December 1991 and on 6 and 13 December 1997.

Feral Canada Geese of the large eastern race *canadensis* have been seen intermittently since 1982 and as numbers increased on various nearby waters, including Southport Marine Lake it was presumably only a matter of time before the species gained a foothold at Marshside. This occurred in 1992 with the sudden arrival of up to 80 birds on M2 in mid-October; some 25 were still present in early December.

First breeding, by one pair on the island in the main Sand Plant lagoon took place in 1993 and three young were reared. Since then the pattern has closely resembled that for the Greylag with flocks feeding on the fresh marsh for a week or more at almost any time of year then abruptly departing only to return after a few days or a few weeks of blessed absence; flocks have on many occasions been watched flighting between Marshside and the Marine Lake. Numbers have increased still further since 1996: 286 were present in September of that year and up to 248 in October 1997. A few hideous hybrids of unimaginable provenance have been seen with the flock since 1993. Breeding has taken place annually since 1995, and in 1998 a record three pairs reared a total of 14 offspring.

BARNACLE GOOSE *Branta leucopsis*

Oakes (1953) cites an assertion that Barnacle Geese had been common on the Ribble estuary in the early 1900s but without providing any evidence. Greenhalgh (1975) discounts the claim, referring to a record of 50 at Crossens on 4 December 1941 as the only documented report of a sizeable flock in South-West Lancashire up to the 1970s. Barnacle Geese in ones or twos were very occasionally recorded with the Pink-footed Goose flocks on the Ribble between about 1950 and 1963. From the mid-1960s records became annual with three or four birds present in most winters into the late 1970s. A steady trend towards larger groups of birds is observable in records from Marshside-Crossens in the 1980s: six were present on 2 April 1980, seven from 7 to 10 March 1982, up to 11 in December 1987, and nine in December 1989. Since then some quite large groups have been recorded: 22 on 2 February 1992, up to 33 in December 1993, 36 in January 1997, and the biggest influx for half a century involving up to 37 birds in January 1999.

Some regular observers have voiced doubts about the status of this increase, casting suspicious eyes at the long-established menagerie of free-flying Barnacles at nearby

Martin Mere WWT reserve. In reply others point to the well-documented growth in Barnacle Goose numbers wintering on the Solway (see Lack, 1986) as the more likely source of at least some of Marshside's geese. This debate is unlikely to be resolved in the foreseeable future and there is probably some merit in both explanations; in December 1999 four birds from a flock of 15 wore Darvic rings, presumably fitted in Svalbard.

BRENT GOOSE *Branta bernicla*

Greenhalgh (1975) reported that the dark-bellied nominate race of the Brent was slightly outnumbered by the pale-bellied race *hrota* on the Ribble in records up to the mid-1970s; both forms were described as very scarce visitors to shore and mosses with no more than one or two recorded in most winters. As far as Marshside is concerned, at least, the picture has changed since then in two significant ways: first, Brent Geese in general have become rather more common and second, the subspecies bernicla is now by far the more numerous of the two, outnumbering the pale-bellied hrota by at least three to one.

Dark-bellied Brents have been recorded in nearly every winter since 1983 with peaks of four in late January 1985, 11 in mid-February 1987, and up to 14 in December 1991. Pale-bellied birds, by contrast, seldom occur in groups; three in February 1988, two from early November 1993 to late January 1994 and three in January 1997 are the only multiple records but single birds have occurred about every other year since 1980. Although Brent Geese at Marshside frequently feed among the grey geese inland of the Marine Drive they tend to favour the outer edge of the saltmarsh or the muddy shore south of the Sand Plant. Records span the period from 7 September to 8 May but the great majority have occurred between November and February.

RUDDY SHELDUCK *Tadorna ferruginea*

There are three documented records: one on 4 November 1990, a female seen on several occasions with Shelducks on the shore south of the Sand Plant from 12 to 17 November 1993, and a bird in flight over Crossens saltmarsh on 9 July 1995. There are three competing opinions regarding the status of this species in Britain and Ireland: vagrant from North Africa or South-West Asia, straggler from a feral population on the near Continent, or escape from captivity. Whichever category or categories our birds belong to (and true vagrants are probably extremely rare in this country) the Marshside regulars are always happy to give them the benefit of the doubt.

SHELDUCK *Tadorna tadorna*

Up to the 1950s the Shelduck bred in fairly small numbers on the Ribble mainly from Longton marsh southwards; it was more numerous on the dunes from Ainsdale to the mouth of the River Alt (Oakes,1953). Since then the population on the sandhills has apparently dwindled but the Shelduck has become an abundant and conspicuous resident of the Ribble estuary. By the mid-1970s up to 1,400 were being counted on Marshside-Crossens-Banks marshes combined in September rising to a peak of 2,000 in November and with as many as 1,100 still present in March. In 1974 nearly 900 birds summered on the Ribble marshes including the mosslands and up to 50 pairs bred on the south side of the estuary alone, mainly on the mosses (Greenhalgh, 1975).

At Marshside-Crossens in particular the species is plentiful at almost any time of year and especially so between October and March when hundreds feed on the muddy shore south of the Sand Plant and on the saltmarsh. Peak totals have remained fairly stable over the past two decades: there were over 500 in January-February 1982, 1,000 on Crossens saltmarsh alone in mid-January 1984, 950 in mid-November 1987, 500 in mid-November 1989, 750 in December 1996 and October 1997, and 820 in January 1998.

Although loafing groups of non-breeding adults and immatures are a common sight on the fresh marshes in late spring and early summer breeding numbers are low. One or two pairs occasionally nest on Crossens saltmarsh and overcome spring tide surges and the numerous predators to lead their clutches onto Crossens channel, joining the creches from Banks marsh in June. Since 1994 two or three pairs have nested on the fresh marsh in most years.

MANDARIN DUCK *Aix galericulata*

An adult male on the M2 creeks on 9 January 1999 was our first record of this feral introduction from East Asia.

WIGEON *Anas penelope*

The Wigeon has been the most numerous wintering duck on the Ribble at least since organised counting began in the early 1960s and probably long before that, although a count of 5,000 on Crossens saltmarsh on 22 October 1941 was considered at the time to be exaggerated (Greenhalgh,1975). The introduction of systematic monthly counts of the whole estuary in the late 1960s established not only that the Banks-

Crossens saltmarsh complex was the Wigeons' favoured feeding area but also that numbers there were increasing dramatically, from 900 in January 1970 to 2,600 in January 1972 and 6,000 in December 1973.

Numbers of Wigeons on the south side of the Ribble estuary continued to grow during the 1970s and early 1980s, no doubt assisted by the establishment of the National Nature Reserve on Banks-Crossens saltmarsh in 1979. The main focus of the flocks, however, seems to have been on Banks marsh proper; numbers on Crossens did not increase at quite the same rate in those years. Totals were impressive, all the same: 2,200 in late December 1979 and 2,740 in December 1982, rising to 5,000-plus in December 1985. From about this time, in the mid-1980s, numbers of wintering Wigeons on the Ribble began to increase at an astronomical rate: by early December 1986 25,000 birds were on Banks-Crossens; estuary totals had passed 80,000 by the early 1990s, and in January 1995 an unprecedented 110,000 were counted. With this spectacular growth the fresh marshes and winter floods at Marshside have come into their own. Small parties of Wigeons had occasionally grazed on the inland side of the Marine Drive for many years but the 1990-91 winter saw the beginnings of a regular and increasing flock which provides a more accessible spectacle than the vast aggregations on the distant and often mist-shrouded saltmarshes. A peak of 400 on M1-M2-Crossens Inner in February 1991 was dwarfed by the 2,200 present in January 1992. In January 1993 numbers reached 2,800, a year later 4,150, in January-February 1995 6,400, in February 1996 8,650, and in February 1998 an amazing 13,400.

Birds begin to arrive in mid-September but in most years it is not until late October that sizeable flocks assemble. After the mid-winter peak substantial numbers remain well into March but by mid-April all but a handful have departed. It is by no means unusual for a few birds to spend part or all of the summer on the fresh marshes but breeding has not so far been suspected.

AMERICAN WIGEON *Anas americana*

A male was found in a remnant flock of Wigeons by Polly's Creek on 24 April 1996 and stayed in the area until 6 May, the first record for Marshside of this long-overdue Nearctic vagrant. The bird was accompanied by a female which most observers considered to be of the same species, although a degree of controversy exists over that identification. A female was seen briefly among the Wigeons on M1 and M2 on 27 November and 25 December 1999.

GADWALL Anas strepera

Until about 1973 the Gadwall appears to have been a scarce autumn passage migrant to the Ribble estuary and the mosslands; after that date a small wintering flock began to build up at the newly-opened Martin Mere WWT reserve (Greenhalgh,1975). During the 1970s and 1980s regular breeding commenced at Martin Mere, at the Mere Sands Wood reserve of the Lancashire Naturalists' Trust and at several other waters in the area; wintering numbers also continued to increase.

At Marshside the first record was of a single bird on 10 April 1980. From then until 1988 the species was an infrequent visitor mainly in winter although a pair spent some weeks on M1 in April-May 1982. In December 1988 five birds marked the beginning of regular wintering by a small flock. Birds typically began to appear in late September, and by the 1990-91 winter the December peak had reached 30, declining to about 12 in 1991-92 but rising again to a high of 38 in January 1993. Since then numbers have fallen considerably, and no more than six were regularly present in the 1996-97, 1997-98 and 1998-99 winters.

Gadwalls normally favour the M1 floods and the creeks on M2. From the outset one or more pairs have tended to linger well into spring; copulation was observed as early as 1989 but there was no proof of breeding until 1995 when two pairs on M2 and another on M1 reared a total of at least 13 offspring. At least one pair succeeded in bringing off a clutch in 1996, in spite of drought, but nesting did not take place in 1997. In 1998 a single pair reared an impressive brood of 12 offspring, and in 1999 our best breeding season to date produced up to three broods from five pairs on the fresh marsh.

TEAL Anas crecca

The Teal is the most abundant winter duck at Marshside after the Wigeon; like that species, numbers have increased greatly on the Ribble estuary as a whole since the 1960s. Greenhalgh (1975) estimated an average of about 3,500 for the whole Ribble area in the mid-1970s of which some 1,400 frequented the saltmarshes. On the basis of monthly count data since the early 1960s he considered the Teal to be primarily an autumn passage migrant with peak totals almost invariably recorded in September-October. This pattern began to change in 1973-74, however, as larger and larger numbers overwintered in the area; the development of the Martin Mere reserve may have played a significant role in this increase.

At Marshside up to 1,500 were present on Crossens saltmarsh alone in December 1979; 1,800 were counted there in December 1981 and over 2,700 were at Marshside-Crossens in December 1983. Teals took to feeding in numbers on the floods and fresh marshes much more quickly than did the Wigeons: 1,630 were counted inland of the Marine Drive as long ago as January 1984. Wintering numbers on the fresh marsh remained fairly stable at a December-January peak of around 1,300-1,600 for the remainder of the 1980s. There were fewer counts on Crossens saltmarsh where numbers are much more difficult to assess accurately, but about 350-500 were present in most mid-winters. In the 1990-91 winter a sudden and rapid increase in numbers on the fresh marsh in particular was noted coinciding with the similar growth in Wigeon numbers there: 2,300 were present in January 1991 and over 3,000 in January 1992. Peak counts during the 1992-93 winter again reached 2,300 in December; 2,400 were present in January 1994 and 2,800 in the following December. Over 3,700 Teals were jostling for space on M1-M2-Crossens Inner together with some 6,400 Wigeons as well as thousands of other wildfowl and waders in January 1995; numbers have remained high in subsequent winters and a record 4,840 were present in December 1998.

Numbers after mid-winter vary considerably from year to year; normally totals drop markedly from mid-February but in some years such as 1985, 1994, 1995 and 1998 a substantial number, up to 2,000 in 1994, are recorded until near the end of March. Marked day-to-day variations in numbers are characteristic of early spring; whether this merely reflects seasonal restlessness among the wintering flock or is due to the arrival and departure of successive waves of migrants is not clear. Recent occurrences of males of the Green-winged Teal among the Teals in March and April, however, suggests that some passage does take place. These readily-identifiable birds are typically seen for a few days in close association with a flock of European Teals and then suddenly disappear; it seems likely that their companion flock has also moved on. In several recent springs up to eight Teals have remained on M1 or M2 until late May. Although there were rumours of nesting by one or more pairs on Crossens during the 1970s there is no evidence of any recent breeding attempt.

GREEN–WINGED TEAL *Anas carolinensis*

This N. American species was first recorded at Marshside on 6 January 1982. There were no further occurrences until 1989, when one fed among European Teals on Crossens Inner from 8 to 16 April. Since then there have been records in most years. One was present on 13 January 1990, another on 25-28 March, and two on various dates from 17 November to the end of the year. In 1991 there were records of a single bird on several dates in January and March involving in total anything between one and four individuals, and another on 7 December. In 1992 one was present on 15 and 29-30 March, and one was on the M1 floods on 12-13 November 1994. One spent over a week on M1 in late March 1995, and one was in front of the Hide on M2 from 12 to 15 January 1998. In 1999 a male spent over a week on the M1 floods in late March/early April, and there were two brief sightings of a bird from the Hide in December.

MALLARD *Anas platyrhynchos*

The Mallard, along with the Starling and a few other species is one of the 'background' birds at Marshside, always present but seldom noticed and not always diligently counted. In part this is due to long-standing doubts about the 'wildness' of many of our Mallards. Greenhalgh (1975) reports that up to 5,000 Mallards were being released annually on the Ribble by wildfowling syndicates in the early 1970s. It is likely that this practice has declined considerably since then but on the other hand there is evidence of regular commuting to and from Marshside by semi-tame stock from the Botanic Gardens, Hesketh Park and other local waters, the same distinctively-marked individuals having occasionally been seen at Marshside on one day and taking bread from strollers in the Park on the next.

Mallards are common at Marshside throughout the year; winter counts, having remained remarkably stable for over a decade declined in the mid-1990s: there were 577 in December 1982, 442 in December 1984, 400 in February 1985, 422 in January 1993, but only 173 in December 1996 and 210 in December 1997. Counts of 390-plus in October and November 1999, however, signalled a return to high numbers at the end of the decade.

Since the mid-1970s, when Greenhalgh estimated that a single pair normally bred on Crossens saltmarsh and two pairs on the fresh marshes, the Mallard has become one of Marshside's most conspicuous nesting species. There were at least 25 pairs inland of the Marine Drive in spring 1995, 22 pairs in 1996, 32 in 1997, an astounding 55 in 1998, and 51 in 1999. Duckling mortality from predation and inclement weather is, however, usually very high.

PINTAIL *Anas acuta*

The Pintail is the only dabbling duck species whose numbers have actually declined on the Ribble estuary as a whole during the past quarter century, although it remains a fairly common winter visitor and passage migrant. Totals increased rapidly in the early 1970s with flocks concentrated on the outer edges of Banks and Crossens saltmarshes and on the muddy shore south to the area occupied by the Sand Plant and the rubble road used by its trucks; 3,700 were present in December 1973 and 4,700 in December 1974 (Greenhalgh, 1975). Numbers remained high until the early 1980s: 2,000 were on Crossens saltmarsh alone in early September 1980, up to 4,000 were there on 11 March 1981, and 2,000 again in late January 1983. Soon after this date, however, Pintail counts began to decline very substantially at the same time as Wigeon and Teal numbers began their spectacular expansion. An annual average of 2,400 wintered on the Ribble as a whole during the early 1990s (Cranswick et al, 1995). 1,000 on Crossens on 9 March 1985 and 750 on 24 December 1991 are the only really large site counts since 1983, although 200 or more may still regularly be seen on Crossens saltmarsh during winter high tides.

Greenhalgh (1975) describes the diet of Pintails on the Ribble estuary as made up mainly of marine invertebrates; it is possible that the slow but steady spread of Spartina cordgrass and other saltmarsh vegetation over the mudflats made the site less attractive to the Pintail in particular though enhancing its value to related species. Although the Pintails in their heyday mainly frequented the shoreline and outer saltmarshes small numbers were also often to be seen on the creeks and floods inland of the Marine Drive: 100 were present on M1 in December 1981, 60 in December 1982 and a then-record count of 197 on M1-M2-Crossens Inner on 26 February 1984. Counts of 50 or more were frequent throughout the rest of the 1980s and up to 167 were present on the fresh marshes combined in January 1989.

Since the 1991-92 winter Pintails have tended to concentrate on the M1 floods; numbers at first increased, from a peak of 220 in March 1992 to 320 in January 1993 and 270 in March 1994, but in recent winters no more than 50 have been regularly present. All Pintails have normally left Marshside by the middle of April; there are a few scattered records of late stayers or passage birds as late as mid-June and in 1998 a pair summered on M2, but breeding has never been suspected.

GARGANEY *Anas querquedula*

Although very rarely recorded on the Ribble up to the 1950s (Oakes,1953) the Garganey was described by Greenhalgh (1975) as a scarce but annual passage migrant since the early 1960s with records about evenly divided between spring and autumn. At Marshside the beautiful drake in particular is looked for with great eagerness on the M1 floods and M2 creeks from mid-March onwards. Although in the 1970s and 1980s quite a few years passed without a record, Garganeys have become much more regular since 1989 with visits in every spring except 1990. The Garganey is often among the first spring migrants, and is all the more welcome for that; the earliest record is of a male on 13 March 1977 and there were late March sightings in 1991 and 1993. Most occurrences have been in the period mid-April to early May with one on 13 June 1993 the latest on record; in 1998, however, three males were present together on M2 from 12 to 17 May, and presumably another three were seen on 8 June. This pattern was repeated in 1999, when up to four males and a female were on M2 between 3 and 19 May. It is surprising that only four birds have been recorded in autumn, in August 1997, July/August 1998, and in August and September 1999.

In late May 1992, after a pair had been present on M1 since 19 April a nest containing at least six eggs was discovered below the Marine Drive embankment. This was the first confirmed instance of breeding by Garganeys on the Ribble estuary. Two small ducklings were observed on 3 June but it was considered unlikely that they could have survived the very rapid drying out of the marsh that ensued; at any rate there were no further sightings.

SHOVELER *Anas clypeata*

Writing in the year 2000, it is startling to be reminded that as recently as 1978 a sighting of even a lone Shoveler at Marshside was considered worthy of comment in the day's notes. Greenhalgh (1975) described the species as a scarce resident and passage migrant on the Ribble with counts in excess of ten being exceptional; he cited comments by a late nineteenth-century source as evidence that Shovelers had formerly been fairly common in the area.

The recent increase in numbers at Marshside began in 1979 when up to 14 were present on the fresh marshes in April and May and two pairs remained into June. Up to 30 were at the site in March 1980 and for the first time a sizeable gathering was recorded in mid-winter, peaking at 23 in late December.

By the mid-1980s a regular winter flock of around 50 birds had become established on M1-M2 with counts sometimes reaching 100 or even 120. These peaks typically occurred in February or early March suggesting that numbers were being augmented temporarily by birds on passage. The situation has remained more or less stable until the mid-1990s, with perhaps a slight further increase in 1993-95; peak numbers in 1996-98, however, were appreciably lower, averaging around 50 in mid-winter and in March. A substantial recovery in 1999 produced our highest-ever counts: 146 in February and 170 in October.

From the very beginning of the period of expansion individuals and pairs have tended to remain into the early summer, and intermittent attempts at breeding have taken place with a fair degree of success below the Marine Drive embankments on both M1 and M2. Since 1983 when a single pair reared at least five offspring, successful breeding by at least one pair has occurred in 1986, 1988, 1991, 1993, and from 1995 to 1999. In 1995 at least four and probably six pairs were involved and a total of about ten ducklings reared; in 1997 at least seven pairs nested, all but one successfully. In 1998 our best nesting season yet saw six broods raised out of a total of 15 breeding pairs. At least three broods were raised in 1999.

POCHARD *Aythya ferina*

The Pochard is an uncommon visitor to the Sand Plant lagoons and adjacent creeks, more rarely to the M1 floods. Most records have been in late autumn and winter but one was present on the lagoons on 9 July 1994 and two in June 1999. Occurrences have become more frequent since the early 1990s and it is likely that most are associated with the recently-established wintering flock on nearby Southport Marine Lake. In 1995 one or two were present intermittently on the lagoons in January, increasing to six by 12 February and to nine by the end of the month; this little flock had dispersed by early March. There were peaks of 12 in 1996, 13 in 1997 and seven in 1999, all in January. Marshside is hardly ideal habitat for a diving duck that typically prefers fairly large bodies of fresh water and there has never been any sign of breeding activity.

TUFTED DUCK *Aythya fuligula*

The pattern of occurrences for this species is broadly similar to that for its close relative the Pochard, but there are two noteworthy differences: first, Tufted Ducks are appreciably the more common of the two and second, records tend to be clustered in autumn and spring although there are sightings in all months. Like the Pochard, Tufted Ducks are seen on the

Sand Plant lagoons, on Polly's Creek and occasionally on the M1 floods; visits tend to be sporadic and unpredictable. As in the case of the previous species sightings have become more frequent in the 1990s and numbers of birds involved have also increased, from singles or occasional pairs prior to 1992 to four in April and nine in September 1993, four again in April 1994 and five in October, 13 in late March 1995, five in March 1996, 18 in October 1997, 13 in December 1998 and 12 in March 1999.

In May 1998, after up to eight apparently-mated pairs had been present since early April, at least one pair was seen with well-grown young on M1; this was the first recorded breeding by this species at Marshside. One pair again nested in 1999 and four young were reared.

SCAUP *Aythya marila*

Oakes (1953) describes the Scaup as occurring quite commonly in winter on the Lancashire coast in general and off the Ribble estuary in particular, with frequent visits by small parties to Southport Marine Lake. By the mid-1970s, however, numbers offshore had greatly declined and only occasional individuals were turning up on the Marine Lake; four there in late January 1975 was the largest number recorded for a decade, according to Greenhalgh (1975).

At Marshside the Scaup remains a fairly scarce visitor mainly between August and April, with only about 20 records of singles or small parties since the late 1970s. Most are seen on the tide but there have been quite a few visits to the Sand Plant lagoons, and this trend seems to be increasing; there were three single birds in 1998, in March, July and August, and two on 25 December 1999. A raft of over 200 on the estuary on 8 January 1997 was quite exceptional in recent decades.

EIDER *Somateria mollissima*

The first record of Eider in the Ribble estuary was of a corpse at Fairhaven in January 1958. Four were off St. Annes in 1960 and from 1967 to about 1974 small numbers were present on the estuary during most winters (Greenhalgh,1975). Occurrences subsequently became much less frequent although there were a few sightings in spring and early summer on Banks marsh in the late 1970s and early 1980s. These visits in the breeding season seem to have persisted into the 1990s as successful nesting by up to two pairs was reported in 1995 and 1996.

At Marshside there have been eight records and these have become more frequent since the mid-1990s: a female was off Crossens saltmarsh on 15 July 1977, a pair was in the same area on 3 January 1983, a flock of seven was on the tide off the Sand Plant peninsula in early May 1985, seven males and three females were on the estuary on 11 September 1992 and an immature male on 23 November 1995; up to three were seen off the peninsula and on Crossens channel on several dates in January 1997, and a female on the M1 floods from 19 to 21 April 1998 was the first record for the inland side of the Marine Drive embankment. If the breeding foothold on Banks marsh becomes established more frequent sightings at Marshside may be anticipated in the future.

HARLEQUIN *Histrionicus histrionicus*

A male was shot by a Mr.Valender at Crossens channel in January 1916 or January 1917 (Greenhalgh,1975). Considering the season and the fact that the species must have been exceedingly rare in wildfowl collections in those days, especially in a free-flying state, it is highly likely to have been a wild bird.

LONG-TAILED DUCK *Clangula hyemalis*

According to Oakes (1953) the Long-tailed Duck was a very rare winter visitor to the Lancashire coast in the first half of the 20th century; among only seven records he cites was one of two birds off Crossens on 25 October 1941. During the 1960s and early 1970s a considerable increase was noted with frequent sightings off Formby

Point and the Fylde coast and regular visits to Southport Marine Lake by up to four birds at a time. Records were widely distributed across the year with most in November but a secondary peak in April may have reflected passage movements (Greenhalgh, 1975). Since the early 1980s the species has again become scarce; it is now seldom recorded on seawatches and there have been only two individuals on the Marine Lake during the 1990s.

At Marshside, apart from the 1941 occurrence there have been two records involving long-staying individuals on the M1 floods: a first-winter male was present from 5 November 1982 to 29 January 1983 and a female was on the residual pool at the corner of Marshside Road and the Marine Drive from 27 April to 8 May 1983. The only other record is of a single bird offshore on 30 October 1991.

COMMON SCOTER *Melanitta nigra*

Although this duck is still recorded in some numbers off Formby Point and the Fylde coast on spring and autumn passage and to a lesser extent in winter, counts are but a shadow of the 10,000 seen off Formby in October 1939 or even of the 3,200 there in late August 1967 (Greenhalgh,1975). In the late 1990s a seawatch count of 500 Common Scoters at the height of autumn migration in September would be considered very noteworthy.

At Marshside the Common Scoter is occasionally recorded in ones or twos on high tides between August and April. Sightings have grown more infrequent since the early 1980s and a group of nine birds offshore on 10 October 1976, up to 25 on 11 September 1992 and 15 on the tide on 1 January 1998 are the highest totals recorded in recent decades. There are a few records of birds on the Sand Plant lagoons and adjacent creeks in winter, a male on 16 February 1999 being the most recent.

VELVET SCOTER *Melanitta fusca*

Velvet Scoters in small numbers were fairly regularly seen off Formby Point in late autumn and winter from the mid-1950s to the early 1970s (Greenhalgh 1975), but the species has become very scarce offshore since then. The only documented record for Marshside is of a male on the tide off the Sand Plant peninsula on 29 December 1978.

GOLDENEYE *Bucephala clangula*

Up to the early 1980s Goldeneyes were fairly common on the sea in winter off both Ainsdale-Formby and the Fylde, and up to ten birds could be seen on almost any visit to Southport Marine Lake between November and February. At Marshside birds were quite frequently seen offshore or on Crossens channel with seven off the Sand Plant peninsula on 3 February 1980 the highest total recorded.

By the mid-1980s a general decline in numbers had set in; few are now seen on winter seawatches and visits to the Marine Lake may have become less frequent. There were no records at Marshside between 1983 and 1987; since 1988 the species has been seen in ones and twos on about a dozen occasions between late October and early April, mainly on the Sand Plant lagoons or the M1 floods, with only two sightings offshore, one on 31 March 1990 and five on 1 January 1998. Goldeneyes have become established as feral breeders in small numbers at one or more mossland sites since the late 1980s. It is not known to what extent the recent pattern of visits to Marshside is attributable to these birds.

RED-BREASTED MERGANSER *Mergus serrator*

This species seems to have decreased in numbers since the late 1960s and early 1970s. Greenhalgh (1975) described it as quite common on the outer estuary between late September and early April, and with up to five together on Southport Marine Lake in several winters prior to 1975. Red-breasted Mergansers appear always to have been scarce at Marshside. There are occasional records of single birds offshore or on the Sand Plant lagoons between late October and late March, but two on the tide on 23 October 1976, a remarkable 17 there on 22 November 1995, and two over the tide in a storm on 25 December 1999 are the only multiple occurrences in the last quarter century.

GOOSANDER *Mergus merganser*

There are only three occurrences of Goosander on record for Marshside: three were on Crossens channel on 3 April 1983 and one flew past the Sand Plant peninsula on 15 May of the same year, and one flew south over M2 and M1 on 20 November 1998.

RUDDY DUCK *Oxyura jamaicensis*

Prior to 1998 there had been only one record of this introduced species, which has become a widespread breeder on lakes and ponds in Lancashire since the mid-1980s, a single bird on Polly's Creek on 21 September 1987. Between 9 May and 2 August 1998 there were repeated and prolonged visits to the Sand Plant lagoons and M2 creeks by up to four males and one female; a good deal of inter-male display took place, but the birds departed before any outcome could be observed. In 1999 up to four males were present on many dates between 9 May and 9 July; some kind of post-breeding pilgrimage to Marshside may be in process of establishment.

HONEY BUZZARD *Pernis apivorus*

A juvenile flew in from the estuary and soared briefly over M2 on 2 October 1998 before moving off south-eastwards; this was the first record for Marshside.

RED KITE *Milvus milvus*

This most elegant of raptors has been recorded three times at Marshside: one flew over the saltmarsh on 12 April 1979, one moved eastwards on 15 January 1994, and a third flew north-eastwards over the main car park on 7 March 1996. With the recently-reported successes of reintroduction programmes in the Midlands it may not be too optimistic to anticipate further sightings in the near future.

MARSH HARRIER *Circus aeruginosus*

The Marsh Harrier is a scarce but annual passage migrant in spring and autumn with occasional visits by wandering immatures in the summer months. Sightings are almost invariably over the saltmarshes but birds are not infrequently seen hunting quite close to the Marine Drive. Most records are in spring and although a few have been seen in March with one on 25 March 1979 the earliest since the mid-1970s, occurrences tend to be concentrated in the period from about 20 April to the end of May. The great majority of spring migrants pass through quickly, often taking only a few minutes to flap and glide over Crossens saltmarsh and away towards the north or north-east. In a typical year two or three birds are seen on spring passage but it is likely that several pass unrecorded in most seasons. In some years one or more immature Marsh Harriers are seen in late June and July. These apparently wandering birds typically stay around the Crossens and Banks saltmarshes for a few days or longer, and also turn up intermittently at Martin Mere WWT reserve during the mid-summer period.

Autumn migrants are seen from early August to late September, more rarely to late October. There are fewer records than in spring but unlike the spring birds autumn Marsh Harriers are often disposed to linger, occasionally for a lengthy period as in the case of an immature bird present on Crossens-Banks from 1 September to 10 October 1976; in autumn 1997 three long-staying visitors each spent an average of 12 days in the area between 20 August and 21 November. Examination of the pattern of occurrences over the past quarter century does not reveal any consistent changes either in dates of records or in numbers recorded, although a female over the saltings on 22 January 1995 was wholly exceptional.

HEN HARRIER *Circus cyaneus*

Although occasional Hen Harriers are seen on spring and autumn passage at Marshside it is as a winter visitor that the species is viewed with most affection by the birding regulars; few sights are more evocative than an adult male Hen Harrier sailing and tilting low over the saltmarsh on a bleak December afternoon. Unfortunately this spectacle, once a regular feature of winter birding sessions, has become more and more rare in recent years as sightings of this species at Marshside-Crossens have dwindled quite dramatically. Oakes(1953) describes the Hen Harrier as a former breeding species on the Lancashire uplands which had declined in status to a scarce passage migrant. By the late 1960s the species's fortunes as a breeder had greatly improved in many parts of Britain and Ireland and wintering had become regular on the Ribble estuary. From the mid-1970s until about 1987 numbers of Hen Harriers in winter at Marshside and Crossens remained remarkably stable with up to three males and four females or immatures seen on a more-or-less daily basis from mid-November until late February. Most sightings were of birds hunting over the saltmarsh but occasional sallies over Crossens Inner in particular were by no means unusual. In the 1985-86 winter a small roost was established for a time on the saltmarsh just north of the Sand Plant and up to seven birds were seen in the vicinity on January evenings.

With hindsight a slight decline both in number of birds and frequency of sightings is evident in the records from about the 1988-89 winter, although it was not clearly discerned by birders at the time. By late 1991 the collapse in numbers had become obvious to all. Although Hen Harriers were still seen at Marshside in the late 1990s no more than two individuals are involved in most winters and sightings are sporadic. It is difficult to assess whether the decline is part of a general slump in Hen Harrier numbers or merely a local phenomenon; sightings along the North Ribble shore seem to be somewhat more regular than at Marshside. While Hen Harriers are still widely recorded from mossland sites in winter most birders are of the opinion that a significant decline across the whole South-West Lancashire area has taken place since the late 1980s. A marked increase in the frequency of sightings occurred in autumn and early winter 1998, and this was maintained in 1999; one lives in hope. We know a little about the origins of our visitors, as individuals from the Islay wing-tagging scheme have been recorded in 1991 and 1993, a bird from a similar project in Perthshire in 1995, and a first-winter male reared in Bowland, Lancashire was seen on several occasions in November 1998.

In addition to occurrences in winter the Hen Harrier is a passage migrant in very small numbers with occasional records in April, May and September.

MONTAGU'S HARRIER *Circus pygargus*

There are three records of this species which has probably always been a rare visitor to North-West England. An adult male was over Crossens saltmarsh on 30 July 1972, another gave some excellent views as it ranged over the area between the Sand Plant and Crossens channel on 12-13 June 1992, and an immature male was over the saltmarsh intermittently on 15-16 May 1997.

GOSHAWK *Accipiter gentilis*

Many visitors to Marshside seem to go through a phase of 'seeing Goshawks'. There are, however, only four documented records of this powerful woodland raptor, an adult male on 13 October 1991 and at least three birds in 1992: one, sex unspecified on 8 January; an immature, probably a male, on 3 and 15 October; and a female on 19 November.

SPARROWHAWK *Accipiter nisus*

The recent history of the Sparrowhawk at Marshside as in most parts of Britain and Ireland is one of extraordinary expansion; it is also a lesson in the adaptability of a raptor species. Following a general and in places severe decline in numbers in the 1960s due to the effects of pesticides a slow but steady recovery took place through the 1970s and early 1980s. During this period sightings of Sparrowhawks became reasonably frequent at Marshside, mainly on Hesketh Golf Course where small passerines and in particular Starlings were the customary prey.

From the middle of the 1980s as Sparrowhawks became an everyday sight over nearby residential areas and even in Southport town centre, population pressure presumably induced some birds to range out onto the fresh marshes, and the frequency of records as well as the number of individuals involved increased at an accelerating pace. From about 1987 Sparrowhawks at Marshside took to hunting even over the saltmarshes and by the early 1990s they were frequently to be seen chasing waders over the open shore and along the edge of the tide. That these forays are often successful is beyond doubt; prey taken, usually by the more powerful females, has included Snipes flushed and killed on low passes over M1 and M2, Knots and Dunlins plucked out of wheeling clouds of waders in masterly style, and even an adult Coot killed and carried off (for a short distance!) in front of several amazed human witnesses in August 1998. Meanwhile the smaller males have competed, often effectively, with wintering Merlins in attacking finches, Skylarks and pipits.

Sparrowhawks are present at Marshside all year round although numbers are highest in winter when up to four birds may be seen on more-or-less a daily basis. It is not known if there are any passage movements. From 1992 to 1995 a single pair reared at least two young each year on Hesketh Golf Course; nesting resumed in 1999, and at least three young were fledged.

BUZZARD *Buteo buteo*

Before 1996 there had been only three recent records, all of single birds flying over, on 8 March 1987, 1 March 1994, and 3 September 1995. A minor explosion of occurrences followed with five records in 1996, in late January, late February, early May and mid-December; the May record involved two birds moving east together.

A few Buzzards are sighted on the mosslands in most winters and some individuals appear to stay around for several months; it may be significant that most Marshside records fall into the periods when such wintering birds would be expected to be moving into or out of the area. Two birds were seen in 1997, in late January and mid-May, and one in late March 1998; by then the position had been complicated by the presence of at least two resident Buzzards, of unknown origin, in the Churchtown-Hesketh Park area of Southport; sightings of one or other over Hesketh Golf Course in particular became fairly regular in the spring and summer of 1998. A bird flying in high over the estuary and moving inland on 25 September was the only record in 1999.

ROUGH-LEGGED BUZZARD *Buteo lagopus*

There are only two records of this sought-after raptor: one was flushed at close range from Crossens saltmarsh on 26 March 1977 and flew off eastwards; another spent nearly half an hour hunting over the saltmarsh on the afternoon of 27 October 1997, and was seen again briefly two days later.

OSPREY *Pandion haliaetus*

There are six documented records of overflying migrants during the last two decades, all involving single birds: on 2 October 1980, 7 September 1988, 3 April 1992 with a different individual on the following day; 25 April 1997, and 6 September 1998.

KESTREL *Falco tinnunculus*

The Kestrel is the most numerous raptor species at Marshside. Up to half a dozen birds may be seen in a day at any time of year and all habitat types are frequented. Although there are no recent records of breeding within our recording boundaries Kestrels nest quite commonly in nearby residential areas, particularly in church towers and similar structures; there is also a healthy population on the mosslands.

Numbers at Marshside and Crossens are at their highest in late summer and early autumn when in most years up to 20 Kestrels congregate and spend several weeks hunting over the saltmarsh. Both adults and juvenile birds are represented in this post-breeding assembly which presumably attracts birds from a wide area. Peak counts are of 33 in late August 1979, 30 in mid-September 1986, and 26 earlier in the same month in 1994 and 1998. Both small rodents and insects such as Craneflies have been seen to be taken from the saltmarsh vegetation but no systematic study of the flocking phenomenon has yet been carried out. At other times of year the diet of Kestrels at Marshside includes a wide range of unidentifiable invertebrates, occasional small passerines, and a substantial number of Lapwing chicks in spring.

RED-FOOTED FALCON *Falco vespertinus*

There are two records of this attractive little raptor, a summer visitor to Central and Eastern Europe; both were presumably overshoots on spring passage. An adult male was over M2 and Crossens Inner on 10 May 1979 and an immature female was harassed by Lapwings over M1 on 25 May 1992. In neither case did the bird stay for more than a few minutes.

MERLIN *Falco columbarius*

The Merlin is a winter visitor and passage migrant at Marshside. In contrast to that other moorland breeding species, the Hen Harrier, numbers have remained more or less stable over the last quarter century and may even have increased slightly during the 1990s. In a typical year a few migrant Merlins are seen from the beginning of August, occasionally as early as mid-July. Most of the early birds are juveniles and some stay in the area for several weeks. In most years one or two Merlins are seen on various dates in September and early October; it is not clear which individual migrants move on and which remain to winter but it seems likely that the wintering birds are installed by the middle of November.

Normally up to four Merlins, typically an adult male and three females/immatures spend the winter months at Marshside-Crossens. Most sightings are on the saltmarshes or over the shore but birds also frequently hunt over the fresh marsh and may often be seen perched for lengthy periods on fenceposts on M2. Prey taken includes a range of small passerines from Linnets and Skylarks to the occasional Starling, as well as Dunlins snatched from high tide wader roosts and a Lapwing taken by a large female in November 1998. A general decline in wintering passerine numbers over the last decade does not seem to have had an adverse effect on the Merlins.

Our winter Merlins usually seem to depart in March but as in autumn one cannot be sure whether individual sightings in early spring are of regular winterers or of passage migrants. There are records up to mid-April in most recent seasons, with one on 15 May 1996 the latest to date.

HOBBY *Falco subbuteo*

Formerly a very scarce migrant in spring and autumn (Oakes, 1953), the Hobby has been recorded with increasing frequency in Lancashire in late spring and summer during the past ten years or so at the same time as the species's breeding population in southern England and the Midlands has grown dramatically. At Marshside Hobbies are rare visitors and occurrences are almost invariably fleeting as a bird circles high over the saltmarsh or makes a brief dash at hirundines over the Sand Plant lagoons before moving on; it is likely that one or two pass unobserved in most years.

There have been only nine documented occurrences since the early 1970s: an immature over Crossens saltmarsh on 2 November 1973, one in the same area on 14 July 1983, a juvenile over M2 on 23 August 1986, an adult over the saltmarsh on 12 May 1994, another over M1 on 23 April 1995, an adult over M2 on 2 June 1995, and singles on 5 June 1997, 26 April 1998 and 7 July 1998. The pattern since 1994 may indicate a real increase in frequency or may merely be a random fluctuation; time will, no doubt, tell.

PEREGRINE *Falco peregrinus*

Until the 1950s Peregrines were frequently recorded on the Ribble estuary in winter (Oakes,1953). During the 1960s the breeding population throughout Britain and Ireland declined dramatically due to the effects of pesticide residues. The enactment of protection measures led to a slow recovery in the 1970s but as recently as 1982 Peregrines were rather scarce and sporadic winter visitors to the Ribble estuary, and to see one in spectacular hunting stoop over the distant edge of Crossens saltmarsh was an unaccustomed highlight of a cold January birding session. Records became a little more frequent after about 1985 but were still mainly confined to the mid-winter period. By 1991, however, a major increase was obviously underway with several sightings at Marshside in every month except July and August.

In the 1990-91 winter one or two Peregrines began roosting on the massive gas-holder at Blowick, within sight both of Marshside and of the mosslands. This practice had become firmly established by 1994 with up to two birds at a time visiting the structure throughout the year. Intermittent display has been observed in spring since 1993 but there has not so far been any attempt at breeding. The Peregrine is now an almost everyday occurrence at Marshside particularly in winter, though there are sightings in all months. Birds are seen over both salt- and fresh marshes and often sit on stakes or fenceposts for up to half an hour at a time. Hunting Peregrines frequently cause mass panics among the flocks of waders and ducks, and prey taken in recent years has included Wigeon, Teal, Lapwing, Golden Plover, Knot and Woodpigeon. On several occasions a Peregrine has been seen to fly directly to the Blowick gas-holder immediately after feeding on a kill.

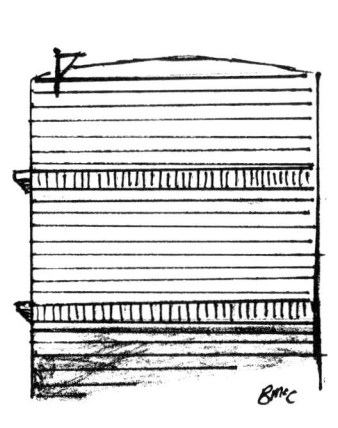

RED-LEGGED PARTRIDGE *Alectoris rufa*

There are two records of Red-legged Partridge at Marshside, a lone bird on the Sand Plant peninsula on 29 March 1989, and another on Crossens Inner on 19 May 1994. A substantial though fluctuating population of Red-legs resides on the mosses, periodically replenished with stock which has in recent years included birds showing some characters of the Chukar, A. chukar. It must be assumed that the Marshside individuals wandered from a nearby mossland source.

GREY PARTRIDGE *Perdix perdix*

As recently as the mid-1970s calm February evenings at Marshside resounded to the hoarse calls of Grey Partridges and displaying birds seemed to be everywhere on the fresh marshes in the dusk, with not a few on the saltings as well. This bird is one of an all-too lengthy list of Marshside breeding species whose numbers have decreased drastically in the last two decades but there is very little data to enable us to trace its decline. Counts are scattered and are mainly of post-breeding coveys in autumn and early winter. Up to 60 were on Crossens Inner and saltmarsh combined in early December 1976, 55 were on M1-M2-Crossens Inner in March 1979, 70 on Crossens Inner alone in early October 1988, and 43 in a single covey on the saltmarsh in late September 1992. Since 1992 no count has exceeded 20 birds.

There are no estimates of breeding numbers extant from the Grey Partridge's heyday but in retrospect at least 20 pairs are likely to have been involved in the mid- to late 1970s. In 1994-96 only two pairs bred on the inland side of the Marine Drive embankment and perhaps another two on the saltmarsh; there may have been a further decline in 1997 and 1998, and by the end of the decade no more than two or three pairs were nesting annually. An accumulation of recent surveys and atlasing projects has identified the Grey Partridge as a species under severe threat throughout Britain and Ireland (see Gibbons, Reid & Chapman,1993). Factors associated with agricultural intensification have been implicated as the most likely causes of this decline. It is difficult to see how such influences could have had a direct effect on the population at Marshside, where land-use changes over the last twenty years have been negligible, unless our birds relied on supplementary food sources on nearby farmland. The reclaimed part of Banks marsh is a possible candidate in this respect.

QUAIL *Coturnix coturnix*

Two Quails were in the sandy south-west corner of M1 on the morning of 10 May 1979. One was heard briefly on 8 May, and another unseen male was calling in vegetation beside Marshside Road on the morning and evening of 8 June 1997. One was calling, and then flushed, from the saltmarsh below the car park early on 6 August 1999. These are the only records.

PHEASANT *Phasianus colchicus*

A small population, almost certainly fewer than ten birds, has been present on Hesketh Golf Course since the spring of 1992. A female was flushed from a nest containing eight eggs in late April 1994, and single females with up to seven small chicks were seen in May 1997 and again in May 1999. Very occasionally a bird wanders out onto M1 and may even venture as far as the Marine Drive embankment. Whether these Pheasants were deliberately introduced or spread naturally from the large mossland population is not known.

WATER RAIL *Rallus aquaticus*

The status of this enigmatic skulker is as difficult to establish at Marshside as it is anywhere else that the species is found. Its retiring habits make for few sightings and even when the species's unmistakable squeals and groans can be heard it is difficult to be sure just how many Water Rails are present. What is clear is that the vast majority of records are in the period from late October to mid-March and that there are far more occurrences in some years than in others. Birds are typically most visible in very cold weather when most or all of the creeks and pools are frozen over. Whether these birds had been present all along and are now being forced to forage in the open, or whether an influx of Water Rails from even colder inland areas is involved, is unknown.

Water Rails are most often seen, or rather glimpsed, in the swampy vegetation at the foot of the embankment on the inland side of the Marine Drive or in the SSSI ditch. The construction of the Hide on M2 has almost certainly contributed to an increase in the frequency of sightings since 1997. Birds are also regularly present on the saltmarsh immediately north of the Sand Plant from which two or three are occasionally flushed by storm-driven tides in winter. Any confident assessment of numbers would be foolhardy but there have been up to three or four different birds seen in most winters since the late 1970s. During periodic influxes or spells of hard weather numbers have been appreciably higher: Up to six were seen regularly in freezing conditions in February 1979 and in February 1994. Five in mild weather on 18 March 1981 may have been migrants. At least five were present and occasionally calling on many dates from 26 November 1994 into early 1995, again in predominantly mild conditions.

There are a few records in spring: one remained from winter until at least 5 May 1982 and one or two birds from the November 1994 influx were occasionally seen until mid-April 1995. Birds were again glimpsed and occasionally heard in dense vegetation near the main M2 creek in June and early July 1995. Some very intriguing piping and squeaking noises were noted, and although there were no sightings of young birds it seems very probable that breeding took place, the first time this has been recorded at Marshside.

SPOTTED CRAKE *Porzana porzana*

According to Oakes (1953) the Spotted Crake was an uncommon but fairly regular autumn passage migrant in Lancashire until the late 1940s. Since then records have dwindled in spite of greatly increased observer activity. One was at Marshside on 12 October 1972 and there were several sightings of a bird in ditches on Crossens Inner near the Water Treatment Plant boundary in January and February 1976. The only other documented records are of one seen briefly on the M1 floods on 24 March 1980, and of an adult which gave excellent if intermittent views below the Marine Drive embankment on Crossens Inner from 31 October to 9 November 1997.

MOORHEN *Gallinula chloropus*

The Moorhen has increased greatly in numbers at Marshside since the mid-1970s. It is now a common and conspicuous resident, present on all the pools and creeks on the fresh marsh and on Hesketh Golf Course. Numbers are occasionally augmented by hard-weather influxes from other sites. From 1976 to 1980 no more than one or two pairs bred in each season and fewer than 15 birds were recorded on the floods and creeks in winter. By 1985 at least six pairs were nesting and up to 40 birds were to be seen on a typical winter's day. In 1992 12 pairs bred and 102 birds were counted on M1-M2-Crossens Inner combined in freezing weather on 21 December.

The population has continued to grow since then with 15 pairs in 1993, 25 in 1994 and at least 20 pairs each year from 1995 to 1999, including up to six on the Golf Course. Although predation and bad weather take a toll of Moorhen chicks, productivity is generally high and the birds appear to cope well with unpredictable water levels and the rapid drying out of the marsh in some seasons. Successful late clutches, either second broods or replacements, seem to be common.

COOT *Fulica atra*

Prior to 1983 the Coot was a winter visitor in varying numbers to the Sand Plant lagoons and to the floods and creeks on M1 and M2; up to 22 in February 1979 was a large enough total to be noteworthy. Numbers tended to increase slightly in the early 1980s; 35 Coots were mainly concentrated on the M1 floods in January 1982 but the birds still abandoned the site in early spring, presumably to breed elsewhere. In March 1983, after some 25 Coots had wintered at Marshside widespread nuptual display, quarrelling and mating took place and eight pairs nested, most of them on M1. Unfortunately only four nests produced young and of these very few chicks reached adulthood as the nests and broods were left high and dry by dramatically falling water levels on M1 in April and May. From 1984 to 1988, although up to 30 Coots were present at Marshside during the winter months breeding attempts were few and sporadic and no more than two pairs nested in any season. In 1989 a massive influx in March brought over 60 birds onto the floods and creeks resulting in nesting by 26 pairs, almost all on M1. Disaster once again ensued as M1 was pumped dry and only two pairs managed to produce any offspring. No breeding attempts were made in 1990, and in 1991 two pairs out of an original eight reared young.

In 1992, however, everything changed with the cessation of spring pumping on M1, and 19 pairs nested with considerable success. Sixteen pairs bred in 1993 and for the first time an appreciable proportion, at least a quarter, of nests were on M2. In 1994 21 pairs of Coots bred and at least 28 pairs nested in 1995; many young were reared. Drought conditions in 1996 and 1997 may have contributed to a decrease to about 18 pairs, but numbers recovered again in 1998 when nesting spread to the ponds on Hesketh Golf Course. 1999 saw the highest total of nests to date, with at least 44 pairs. A substantial number of the Coots reared at Marshside appear to remain at the site all year and wintering totals on the floods have increased to 65 in December 1993 and over 80 in February 1995.

CRANE *Grus grus*

An immature flying over Crossens on 7 May 1978 is the only record.

OYSTERCATCHER *Haematopus ostralegus*

The Oystercatcher is one of several important wader species for which the records for Marshside-Crossens specifically are rather patchy and inadequate. A great deal of data is available for the Ribble estuary overall but most of it does not distinguish between Crossens and Banks marshes, in particular, or between the shoreline north and south of our area boundary at the Marine Drive-Hesketh Road junction. Though inconvenient for our purposes this is not important in itself as in conservation terms it is the Ribble estuary as a whole that is the crucial ecosystem, and an average of 16,500 were present in the area each winter during the early 1990s (Cranswick et al 1995).

The Oystercatcher is obviously common and conspicuous at Marshside-Crossens throughout the year with highest counts in the autumn and winter months: 2,600 were on Crossens saltmarsh on 8 February 1977 and 3,600 there on 6 December 1979. 1,370 were roosting on the shore south of the Sand Plant on 10 January 1982 and an exceptional 7,000 gathered on the tideline between Southport pier and Crossens channel on 12 December of the same year. There are few site-specific counts since the early 1980s but casual observation suggests that numbers have changed little over the years; if anything a slight overall increase may have occurred. Oystercatchers normally feed far out on Horse Bank and on the other mudflats and sandbanks bordering the Ribble channel. At high tide they tend to roost in dense flocks along the water's edge and seldom approach close to the Marine Drive. On the highest of tides the roosts usually decamp en masse to the exposed portions of Crossens saltmarsh, or gather on the fresh marshes, to await the ebb. One or two pairs of Oystercatchers have nested on M2 and Crossens Inner in most recent springs.

BLACK-WINGED STILT *Himantopus himantopus*

One was at Polly's Creek briefly on the morning of 28 April 1984 before flying off over the saltmarsh. This is our only record of this elegant southern wader.

AVOCET *Recurvirostra avosetta*

One on Crossens channel on 9 September 1972 is the only record.

LITTLE RINGED PLOVER *Charadrius dubius*

There were numerous records of Little Ringed Plover in South Lancashire from the late 1960s, some of which involved successful or failed breeding attempts. By the mid-1980s the species had begun to breed regularly in small numbers at several sites including Martin Mere WWT reserve and the Lancashire Trust's reserve at Mere Sands Wood. At Marshside Little Ringed Plovers are scarce but increasingly regular visitors to the fresh marsh on spring and autumn passage. The first record was in spring 1980 followed by a cluster of brief occurrences in 1981-

83, with single adults in April and May 1981, three presumed different juvenile birds in July and August 1981, another adult on 18 April 1982, and three together on 10 July 1983. The species was absent from 1984 to 1988 but there was a pair on M2 from 21 to 29 April 1989 which might have stayed longer but for sustained harassment by a breeding pair of Ringed Plovers.

In 1994 after four more seasons without a record a pair arrived in early May and nested by the Sand Plant lagoons. The nest and eggs were inadvertently destroyed by digging operations only a day or two from hatching and the parent birds were not seen again. Hopes were high of a renewed and this time better-protected attempt in 1995, but records that spring involved only three very brief sightings of migrant birds. In 1996 a bird was present in the lagoons area on 22-23 April, and one or two spent a week in the same location in late July. In 1997 the picture was very similar, with singles in late May and early August. 1998 saw a major surge in passage records, with six birds, including a pair, in spring and at least four birds in autumn; although fewer were recorded in spring 1999 the chances of another nesting attempt in the near future must be good.

RINGED PLOVER *Charadrius hiaticula*

The Ringed Plover is a fairly common passage migrant in spring and autumn and up to very recently a regular breeding species in small numbers. It is rather scarce from late November to mid-January. Ringed Plovers are usually most easily seen in high tide wader roosts on the shore south of the Sand Plant. The birds tend to favour the higher parts of the beach close to the Marine Drive and often form little groups on or beside tussocks of vegetation. There are few detailed counts but at least 200-300 are usually present at peak migration periods in May and August-September; over 600 were counted on 15 May 1988 and 1,200 on 12 August 1999. In some recent springs substantial numbers of migrants have also frequented the remnant floods on M1; between 120 and 150 birds were present on most dates in May 1994, and an exceptional 350 on several dates in May 1998.

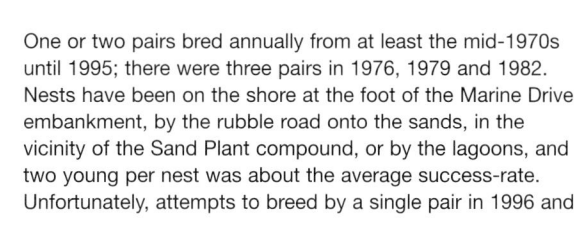

One or two pairs bred annually from at least the mid-1970s until 1995; there were three pairs in 1976, 1979 and 1982. Nests have been on the shore at the foot of the Marine Drive embankment, by the rubble road onto the sands, in the vicinity of the Sand Plant compound, or by the lagoons, and two young per nest was about the average success-rate. Unfortunately, attempts to breed by a single pair in 1996 and

1997 failed, for unknown reasons, and nesting was not even attempted in 1998 or in 1999.

KENTISH PLOVER *Charadrius alexandrinus*

There are four records of single birds on the shore south of the Sand Plant: on 13 October 1977, 17 December 1979, a male on 23 October 1980, and a female on 17 September 1989. In the cases of the two most recent records, at least, the bird was associating with Ringed Plovers in a high tide wader roost. All four visits were of very brief duration.

DOTTEREL *Charadrius morinellus*

For many decades, and possibly for many centuries, small trips of Dotterels have been turning up on passage at traditional sites on the Lancashire mosslands and fells and occasionally on the Ribble estuary. At Marshside there are only three recent records: one was on Crossens from 25 August to 2 September 1973, a group of 13 alighted briefly on M1 on the afternoon of 29 April 1989 before moving on to the east, and a single bird, probably a male, fed with Lapwings on a fairway on Hesketh Golf Course early on 9 May 1993.

AMERICAN GOLDEN PLOVER *Pluvialis dominica*

There are two records of single juveniles among Golden Plovers on M2-Crossens Inner in autumn, on 12-13 November 1984 and on 27-28 October and again briefly on 9 November 1998.

GOLDEN PLOVER *Pluvialis apricaria*

The Golden Plover is an abundant passage migrant and winter visitor at Marshside; numbers in winter increased substantially during most of the 1990s. Autumn migrants begin to appear on M1 or on the shore south of the Sand Plant as early as mid-July in some years although numbers rarely exceed 500 before the middle of September. The great majority of these early Golden Plovers are adult birds moulting out of breeding plumage. After the beginning of October numbers increase rapidly. It is assumed that most of these later birds are destined to overwinter, a hypothesis supported by observations of a small number of distinctive albinistic or leucistic individuals over the years. Numbers peak between mid-December and late January but remain at high levels well into March.

The pattern varies somewhat from year to year, as hard weather causes the flocks to vacate the site for the duration although the birds normally return quite quickly following the thaw. Flocks tend to concentrate on M2 and Crossens Inner with satellite groupings on M1 and on the grazed portion of the saltmarsh. There is usually quite a lot of moving about in the course of a short winter's day, often due to the activities of Peregrines and other raptors. Until the mid-1980s 1,500 to 2,300 Golden Plovers normally wintered at Marshside; 3,800 in January 1978 and 2,660 in late December 1984 were the highest counts recorded up to 1987. There has been a fairly steady increase in winter peaks since then: 4,100 in mid-December 1988, 5,000 in mid-January 1990, 4,800 in December 1992, 6,220 in February 1993, 5,800 in December 1993, 5,200 in December 1994, 5,680 in December 1996 and 6,200 in December 1998.

After remaining high until early March, when up to 3,500 were present in 1995 and 1996, numbers of Golden Plovers decline rapidly by mid-month. Many have acquired their stylish breeding plumage by this time. About 1,500 are typically counted in mid-April but the picture may be complicated at this time by the arrival and departure of migrant flocks. By the end of April all have normally gone, although there are occasional records of stragglers until mid-May.

GREY PLOVER *Pluvialis squatarola*

The Grey Plover is a common shorebird at Marshside both in its spectacular silver-and-black summer plumage and in its drab grey non-breeding dress. The species may be seen all year round although numbers are highest in winter and during the late spring and autumn passage periods. Grey Plovers normally congregate in high tide roosts with other waders on the shore south of the Sand Plant, though a few occasionally take refuge on the fresh marsh during stormy weather. There is no evidence of any consistent change in numbers over the past 20 years. Around 600-700 are usually present in mid-winter but 1,230 were counted on 20 December 1987 and 2,510 on 27 December 1984 was wholly exceptional. Numbers seem to decrease somewhat in late winter and early spring only to build up again with the main spring passage in April and May: 648 were present on 17 April 1983, 1,175 on 13 May 1984, and 1,000 on 8 April 1989.

Relatively few are seen in mid-summer but by early August returning adults, many still resplendent in breeding plumage, begin to gather on the shoreline: 200 on 23 August 1982, 350 on 12 August 1983 and 270 on 31 August 1993 are typical totals. There is a steady increase in numbers both of adults and of juveniles through the autumn with counts of 400 to 700 normal between September and November, but up to 1,100 were present on the shore on 25 September 1999.

LAPWING *Vanellus vanellus*

The Lapwing is among the most characteristic birds of Marshside. A common breeding species and an abundant passage and winter visitor, Lapwings are almost always present and usually highly conspicuous. During the early 1990s an annual average of over 25,000 Lapwings wintered on the Ribble as a whole (Cranswick et al 1995). Noisy territorial displays begin in mid-February if the weather is mild, even while large flocks of wintering Lapwings still remain; the great majority of nesting territories are on M1-M2-Crossens Inner although a scattering of pairs also nest on the saltmarsh.

Breeding numbers remained high from the late 1970s to the late 1990s: there were 186 pairs on the fresh marsh in 1979, 210 pairs in 1980, 137 in 1982, 210 pairs once again in 1993, at least 105 pairs in 1995, 119 in 1996 and 145 in 1997. No more than 90 pairs nested in 1998, however, and the total dropped even further to 67 in 1999.

Although the time of peak nesting activity varies from year to year with weather conditions and especially with flood levels on the fresh marsh, many birds are sitting by early April and the first chicks are often seen by the middle of the month. Pairs continue to lay, or re-lay, well into May, however, and a definitive count of the breeding population is difficult to arrive at since new nesters and clutches of chicks of diverse ages are usually present simultaneously on the same patch of ground. The parent birds are vigilant and demonstrative in defence of their offspring but in spite of their efforts casualties among the newly-hatched chicks are heavy and it is unusual to see more than two half-grown youngsters accompanying any pair. Sparrowhawks and above all Kestrels are the most conspicuous hazards in the lives of the young Lapwings but Grey Herons and the larger gull species also account for quite a few, while an unknown proportion falls victim to the more stealthy Weasels and Brown Rats as well as to climatic factors such as cold, damp and drought.

By the middle of June post-breeding flocks begin to build up, dividing their time between the shore and the now largely dry M1 and M2. Both adults and juvenile birds are included and up to 500 are often present in July and August. Numbers increase steadily through the autumn: September counts of 700-1,000 are typical and by the end of October up to 2,000 are normally scattered over the fresh marsh, rising to 3,000-3,500 in

November. It is not known what proportion of the autumn Lapwings are winter visitors and how many are merely passing through but in most seasons numbers tend to stabilise by mid-December and remain largely unchanged until late February, weather permitting. Numbers overwintering vary somewhat from year to year and peak mid-winter counts in recent seasons have included 4,000 in December 1993 and 5,600 in January 1995, 3,000 in January 1996, 6,050 in late November 1997 and 7,300 in December 1998.

The onset of a severe freeze produces an almost immediate desertion of the site by the Lapwing flocks but as in the case of the Golden Plover numbers recover again quickly following the thaw. In early spring as prospective nesters swoop and call around them the winter flocks dwindle rapidly, from 1,800-plus in early March virtually to zero by the month's end.

A substantial recent decline in the breeding population of Lapwings across large areas of Britain and Ireland has been well documented (see Gibbons, Reid & Chapman, 1993). Although the numbers on passage, wintering and nesting at Marshside have remained reasonably buoyant during this period, perhaps the sudden fall in the breeding population at the century's end is a portent of harder times to come.

KNOT *Calidris canutus*

This unobtrusive and usually drab-plumaged wader is, in fact, Marshside's most abundant bird, forming vast roosts at high tide on the shore south of the Sand Plant or on Crossens saltmarsh. Even in the wildest of conditions very few seek refuge inland of the Marine Drive. Numbers are at their highest during the spring and autumn migration periods when counts of over 10,000 are quite normal, 15,000-plus by no means unusual, and 50,000 on 6 September 1978, 54,000 on 12 August 1984, and 56,000 on 4 May 1985 are the highest totals recorded since the early 1970s, at least.

Numbers of Knots in winter and early spring are rather lower than the autumn peaks; 5,000 to 8,000 is typical but 15,000 were present on 10 February 1974, up to 30,000 on 8 March 1981, and 17,600 in January 1998. Knot counts are usually at their lowest in mid-June when around 2,000-3,000 are typically to be seen on the shoreline over the high tide. There may have been some decrease in overall numbers of Knots at Marshside during the last two decades. It is certainly the case that none of the published counts from more recent years match the huge totals chalked up prior to the mid-1980s.

SANDERLING Calidris alba

Although these energetic little waders are fairly common along the sandy shore between Southport Pier and the mouth of the River Alt they are seldom present in very large numbers at Marshside. Sanderlings are usually to be found scattered among the other Calidris sandpipers in high tide roosts on the shore south of the Sand Plant. They are winter visitors as well as passage migrants in spring and autumn. Counts for Marshside-Crossens per se are few and far between but 50-100 is a typical mid-winter total while up to 300 are occasionally seen at peak passage times in August-September and in May. An extraordinary 7,000 were on Crossens saltmarsh on 23 May 1971 (Spencer,1973) and 6,700 were there on 29 July 1973, but 1,000 on 14 May and 700 on 27 August 1980, 1,610 on 1 June 1985 and 550 on 15 August 1999 are the highest counts recorded in more recent years. The spread of saltmarsh vegetation southward along the foreshore has probably meant a reduction in the attractiveness of this roosting area for the Sanderling in particular.

LITTLE STINT Calidris minutus

The recent history of the Little Stint at Marshside is complex and interesting. The species is predominantly a passage migrant, occasionally in some numbers, but there is also a pattern of sporadic overwintering by small groups of birds. Little Stints are scarce on spring passage although records have become a little more frequent since the late 1980s. All the more recent sightings, at least, have been in late April or May on the remnant M1 floods or in the Sand Plant lagoons-M2 creeks area. Most spring Little Stints are in breeding dress although a few retain traces of winter plumage; three on 3 May 1981 and four from 18 to 22 May 1994 are the highest numbers recorded.

Autumn Little Stints are usually seen among other Calidris waders in high tide roosts on the shore south of the Sand Plant. Numbers vary considerably from year to year; there were none at all from 1985 to 1987 and in 1989-90, but up to 50 were at Marshside in autumn 1969, 42 were at Crossens on 27 September 1973, 35 were present in mid-October 1976, 18 on 28 August 1980, up to 10 on several dates in September 1982, six on 8 September 1991, and eight on 10 September 1995. Autumn 1996 was remarkable for two distinct waves of migration, the first from 15 to 28 September involving at least 20 birds, and the second between 13 and 16 October which involved at least 35 birds. Peak counts during the three autumns from 1997 to 1999 averaged 16 birds. Autumn records extend from late July to mid-November but the majority of Little Stints are seen in the period from late August to late September when almost all are in fresh and distinctive juvenile plumage.

Overwintering by small parties of Little Stints has taken place many times in the last quarter century. There is, interestingly, no relationship between numbers present during the main autumn passage period and the occurrence of birds in winter. The fresh marshes and in particular the M1 floods are the preferred haunt of Little Stints in winter: up to five were present each winter between 1976 and 1979, dropping to two in 1980. Two were present in December 1985-January 1986 and three in winter 1987-88. There were no further winter records until late January 1994 when two birds appeared on the M1 floods, increasing to five from early March to 10 April. One of the 1994 Little Stints remained on M1 until 8 May; this was unusual, as in previous years overwintering birds had departed by mid-April, at the latest. Two were again present in December 1994 and these frequented the M1 floods until early April 1995; three on the shore on 11 December 1996 were not, however, seen again. Three birds which appeared on M1 on 2 December 1998 remained into 1999. Three were again present in November 1999, and at least one remained into the New Year.

TEMMINCK'S STINT *Calidris temminckii*

There are six records of single birds: on 25 June 1974, 5 May 1976, 21-22 May 1989, 16-18 May 1997, 16-17 May 1998, and 20 May 1999. All the birds since 1989, at least, have been seen on the Sand Plant lagoons. It is possible that the 1997-99 records refer to annual visits by the same individual, stopping off on its journey north.

BAIRD'S SANDPIPER *Calidris bairdii*

There have been two records of this Nearctic vagrant, each involving an autumn juvenile on the shore south of the Sand Plant. The first was present from 19 to 25 September 1982 and the second was in a high tide roost with other Calidris species on 12 September 1991.

PECTORAL SANDPIPER *Calidris melanotos*

A juvenile on M1 and M2 on several dates between 5 and 12 October 1999 was our long-awaited first record of the most frequently-recorded Nearctic wader in Britain and Ireland.

CURLEW SANDPIPER *Calidris ferruginea*

The pattern of occurrences of the Curlew Sandpiper at Marshside parallels quite closely that of the Little Stint in that the present species is also primarily an autumn passage migrant with smaller numbers recorded in spring and with some instances of overwintering during the last two decades. Autumns with high numbers of Little Stints also tend, by and large, to be those in which more Curlew Sandpipers than usual are recorded. The Curlew Sandpiper is a scarce and irregular migrant in spring. There were none at Marshside between 1976 and 1985 and since then only 1986, 1990, 1992 1994, 1996, 1998 and 1999 have produced sightings with four on 25-26 May 1992 the largest number recorded until an exceptional influx of at least 15 on the remnant M1 floods in late April and May 1998, peaking at seven on 14 May.

In some years one, or very occasionally a few, adult birds are seen in late summer. These are assumed to be failed breeders returning south, and six on 30 July 1984 and two on 26 July 1985 are the only multiple occurrences in recent years. Curlew Sandpipers in spring and summer are seen on the remnant floods on M1 or at the Sand Plant lagoons and adjacent creeks. By contrast most occurrences during the main autumn passage period are in high tide roosts on the shore south of the Sand Plant or,more rarely, on Crossens saltmarsh. The great majority of autumn migrants are juvenile birds with immaculate scaly mantle and peachy-buff underparts. They are readily distinguishable among the common herd of Dunlins even at quite long ranges. As in the case of the Little Stint, autumns vary greatly in the numbers of Curlew Sandpipers recorded: 1972, 1974, 1978, 1980, 1982, 1985, 1988, 1991, 1995, 1996 and 1999 all stand out as seasons in which up to 30 Curlew Sandpipers were recorded together, and the 39 seen on the shore and Crossens saltmarsh combined on 5 September 1988 is the highest number ever seen at Marshside. Most autumn migrants pass through between early September and mid-October. Some groups linger for several days or even weeks and there are occasional records in November and early December. In 1979-80 and 1987-88 single Curlew Sandpipers overwintered at Marshside; in each case this occurred after a relatively small total of birds had been recorded in autumn.

PURPLE SANDPIPER *Calidris maritima*

Marshside would appear to be an unlikely choice of habitat for this frequenter of rocky shores and stone jetties, so it is not surprising that the Purple Sandpiper has been recorded on four occasions only: single birds among Turnstones on the rubble road used by the sand trucks on 9 September 1975 and 16

September 1981, and two there on 15 September 1992. One was in a high tide roost on the shore by Hesketh Road on 11 September 1999.

DUNLIN *Calidris alpina*

Although counts of Dunlins over the Ribble estuary as a whole frequently exceed 30,000 during the winter months as well as during passage periods, and totalled 51,415 in November 1994 (Cranswick et al 1995), numbers at Marshside in particular have never quite attained the maximum totals reached by the Knot. The Dunlin is, nevertheless, an abundant wader. It is, in fact, usually rather more conspicuous than its larger cousin since at high tide roosting Dunlins typically congregate near enough to the Marine Drive to allow some extremely close views, while Dunlins also feed and roost quite freely on the fresh marshes, a practice rarely observed among Knots.

Dunlin numbers at Marshside appear to reach their peak in late winter and early spring although this is yet another common species for which detailed site counts are more sparse than one would wish. 52,000 at Crossens on 17 November 1974 is the highest published total in recent decades and other noteworthy counts include 8,400 on 10 January 1982 and 12,000 on 2 February 1998. Substantial flocks pass through on spring and autumn migration: 12,000 on 26 March 1989, 21,000 on 16 September 1997, and 11,577 on 18 March and 13,000 on 12 August 1999 are the highest numbers recorded in recent years.

BROAD-BILLED SANDPIPER *Limicola falcinellus*

There are three records: one on Crossens saltmarsh on 16 October 1954, another in the same area on 1-4 July 1976, and a third in a high tide wader roost on the shore south of the Sand Plant on 5 May 1988.

RUFF *Philomachus pugnax*

From being a scarce passage migrant at Marshside up to the late 1970s the Ruff had become one of our most noteworthy birds by the mid-1990s, occupying the multiple roles of migrant, summer resident and probable breeder, and winter visitor. Four Ruffs on M2 on 29 September 1976 and one on M1 on 17 May 1977 were considered to merit special comment at the time but by the spring of 1978 up to 19 birds spent several weeks on the fresh marsh. 17 were there in May 1979 and two overwintered in 1979-80. There were no further records in 1980 and 1981 but lekking by at least four birds was observed on M1 and M2 in mid-May 1982 and 46 were present at the year's end.

This set the broad pattern for the rest of the 1980s: variable numbers, usually including male birds in breeding plumage, turned up in late spring. Lekking was recorded in 1986, 1987 and 1989, and copulation was also observed in 1987. Numbers in spring varied considerably from one in 1984 to 22 in 1986 and 38 in 1987. During the same period the Ruff became firmly established as a regular autumn migrant and as a winter resident of the fresh marshes: peak counts include 47 in early February 1985, 57 in October 1987, and an unprecedented 110 on 27 January 1989. Numbers of Ruffs continued to grow steadily during the early and mid-1990s and winter totals reached 117 in January 1993 and 135 in February 1994; after the 1995-96 winter, however, numbers declined dramatically and relatively few were present at any season in 1997 or in 1998. In 1999 peak counts of 68 and 82 during spring and autumn passage, respectively, raised hopes of a return to the days of plenty, but mid-winter numbers remained very low.

Lekking and copulation were observed in May 1992, in May 1994 (when up to seven males and 28 females were involved) and in April-May 1995. In the early and mid-1990s up to a dozen moulting adult Ruffs tended to gather on M2 from about mid-July, suggesting that breeding may have occurred at some nearby site. Prior to 1995 episodes of display at Marshside had always petered out as the birds dispersed before the end of May; in June 1995, however, at least two males and one female remained in the immediate area and it is probable that breeding was at least attempted.

JACK SNIPE *Lymnocryptes minimus*

Jack Snipes are often discussed in the same breath as Water Rails by Marshside regulars, for like the rail the Jack Snipe is an inveterate skulker, rarely seen except when flushed by tide or human disturbance. The two species are also similar in that numbers actually present in any given season are impossible to estimate with any confidence and the timing and extent of migratory and other movements are virtually unknown. What is known is that the Jack Snipe is a winter visitor and passage migrant recorded in very small numbers between mid-October and mid-May. Most occurrences are in the period from early December to the end of February. The great majority of sightings are brief flight views of single birds flushed from the saltmarsh by birders or by a high tide, and the areas immediately north and south of the base of the Sand Plant peninsula have been favoured in the recent past.

One or two Jack Snipes are seen in most winters; six in January-February 1984 and four in early March 1989 are the highest numbers recorded. There has been a decline in the number of winter sightings since about 1991 but it is difficult to assess the significance of this in so scarce and furtive a species. There are a few scattered records of migrant Jack Snipes in spring; one was flushed from the edge of M1 on 26 April 1991 and another spent several days on the remnant M1 floods in mid-May 1994, feeding in the open and even indulging in evening display flights.

SNIPE *Gallinago gallinago*

The Snipe has experienced a clear and indisputable decline in numbers at Marshside over the last two decades; most of this decrease appears to have occurred since about 1990. Snipes are predominantly winter visitors and autumn passage migrants. Numbers were formerly at their highest in mid-winter and although Snipes are typically dispersed widely over both fresh- and saltmarshes so that accurate counting is difficult, some impressive estimates of numbers on M1-M2-Crossens Inner at least are available: up to 370 in January-March 1979, 400 in December 1982, 250 in December 1985 and a phenomenal 654 on 24 November 1989. The drop in numbers of wintering birds after 1990 has been dramatic; peak counts since have included 65 in late January 1991, 18 in December 1992, 35 in February 1993 and 30 in late November 1994.

Counts during the main late autumn passage period have also slumped from 170 on 2 October 1983 and 255 on 30 October 1985 to 40 in late September 1993 and 60-plus on fresh and saltmarshes combined on 12 November 1995. Spring movements of Snipes have always been rather sporadic and 60 on 3 April 1983 and 65 in late April 1993 are the highest totals recorded.

A decline has clearly taken place in the number of Snipes breeding on the fresh marshes, although the picture is obscured by the absence of counts for many seasons: eight drumming males held territory in 1983, 11 in 1992, 12 in 1993, and five in 1994; a mere three drummers were heard in 1995 and four in 1996; in 1997 only a single bird was heard, two at the most in 1998, and two or three territories were held in 1999.

GREAT SNIPE *Gallinago media*

One was observed for 20 minutes 'on a flooded pasture' at Marshside on 19 March 1926 and described in British Birds. Oakes (1953) considered the record to be fully acceptable.

LONG-BILLED DOWITCHER *Limnodromus scolopaceus*

One found on the M1 floods on 19 November 1998 was the first record at Marshside of this Nearctic vagrant. It was seen again in the same area on a number of dates in November and December, and remained into 1999; the last sighting was on 14 May.

WOODCOCK *Scolopax rusticola*

Lone Woodcocks are occasionally flushed from copses on Hesketh Golf Course, especially during spells of hard winter weather. There are only four recent records from the marsh proper: one flushed from the foot of the Marine Drive embankment on 16 April 1980 was presumably on passage, while another in a frozen ditch on M1 on 16 February 1991 may have been a refugee from even colder conditions inland; one flushed from the verge of Marshside Road on 15 January 1998 is more puzzling, given the mild weather generally prevailing at the time. A single bird was on the peninsula on 23 November 1999.

BLACK-TAILED GODWIT *Limosa limosa*

The Black-tailed Godwit is an abundant passage migrant and winter visitor at Marshside. As the flocks usually frequent the fresh marshes, and in particular the area immediately in front of the Sandgrounders Hide, this approachable and elegant wader is familiar even to casual birders. The species has increased enormously on the Ribble estuary as a whole since the early 1950s, when Oakes (1953) could describe a flock of 145 at Lytham on 6 September 1948 as the largest gathering ever recorded in Lancashire. Black-tailed Godwits on the Ribble are characteristically very mobile and unpredictable. They feed and roost on both salt- and fresh marshes and large flocks tend suddenly to arrive at a particular site and then just as abruptly depart, making an assessment of the total population difficult.

An annual average of 930 wintered on the Ribble in the early 1990s, according to Cranswick et al (1995). In view of totals at Marshside alone during those winters, this would appear to be a considerable underestimate. Counts at Marshside increased slowly during the 1960s and 1970s, especially at migration periods. By about 1980 up to 300 were present in mid-April and 800-1,000 in early- to mid-September. Winter totals were typically much lower, around 45- 50 being a normal December peak. In the early 1980s numbers of Black-tails at Marshside increased substantially with autumn and winter counts showing the most spectacular growth. In 1983 flocks of up to 200 were recorded on spring passage in mid-April while 2,180 were counted on 16 October and 1,150 were present at the year's end. During the remainder of the 1980s the species consolidated its position at Marshside. Spring counts reached 400 in early April 1985 and 1,000 at the same time in 1989, while an exceptional autumn passage peak of 2,700 was reached in early October 1987; between 700 and 1,000 birds normally overwintered.

There were various fluctuations in passage and wintering numbers during the 1990s, but an overall pattern of increase is evident; by 1998 peak counts during spring passage had reached 4,000, with rather lower totals in autumn but an almost incredible 4,300 birds overwintering.

Display is frequently observed among late-staying winterers in March as well as among migrant birds in April and early May, many of which are in full breeding dress. Copulation was also recorded on several dates in May 1995. In 1997, 1998 and 1999 up to 400 birds remained in the Marshside area throughout the summer; there has, however, never been any firm evidence of nesting.

BAR-TAILED GODWIT *Limosa lapponica*

Though often many times more numerous than the Black-tail, the Bar-tailed Godwit is rather less familiar to birders at Marshside, as it confines itself almost exclusively to the shore and outer edges of the saltmarsh. Massive and densely-packed roosts assemble there at high tides between August and May. A few Bar-tails occasionally roost, or more rarely even feed, inland of the Marine Drive particularly during winter storms; 890 on M2 and Crossens Inner in a gale on 25 December 1999 was very exceptional. Count data for Marshside per se are far from comprehensive but it seems that numbers tend to peak in early autumn and again in late winter. 10,000 were on Crossens and Banks saltmarshes combined on 8 August 1979, 10,000 on Marshside shore alone on 11 September 1983, and 9,300 there on 11 August 1984. The highest count recorded in recent years was 14,000 on Crossens saltmarsh on 30 January 1983, while over 7,000 were there on 8 February 1982.

Numbers seem to remain high well into March but counts for the late spring passage period are frequently lacking; 500 to 800 have been recorded on high tides in May in 1991, 1992 and 1999. Given the paucity of published data for this species in the 1990s it is not possible to reach even a tentative conclusion regarding any changes in numbers of Bar-tailed Godwits at Marshside.

WHIMBREL *Numenius phaeopus*

Although substantial numbers are frequently recorded both on the Inner Ribble estuary and on Formby Moss, at Marshside the Whimbrel is an uncommon and unpredictable migrant in spring and autumn. The relative brevity of the species's passage periods in conjunction with the tendency of birds to move through quickly, often passing over without even pausing to feed but uttering their beautiful bubbling call, gives the Whimbrel a peculiar appeal for all the Marshside regulars. Over the last quarter century spring migrants have been recorded between 9 April and 30 May but the great majority pass through from about 15 April to 10 May and most are seen on or over the fresh marshes. 47 on 7 May 1991 and 27 on 22 April 1996 are the largest flocks recorded in spring but there were no spring Whimbrels at all reported in 1979 or in 1982 and fewer than three birds in each of the six successive years from 1983 to 1988, and again in 1999.

Returning Whimbrels are seen from about 10 July to the beginning of September. Autumn birds are usually recorded in ones and twos but up to 26 were present at Crossens channel on several dates in July and August 1979 and up to 24 were on the shore on 1-3 September 1993. As in the spring, however, there were few records, or even none at all, in several seasons during the 1980s. There have been several reports of individual Whimbrels well outside the normal migration periods: one on 8 December 1979, another on 4 November 1982 and a third on 31 December 1998 and 1 January 1999 are the most recent occurrences.

CURLEW *Numenius arquata*

Curlews are common winter visitors and passage migrants at Marshside although there are records in all months. Birds feed and roost on the saltings and on the open shore as well as on the fresh marsh but it is only for the inland side of the Marine Drive that any systematic count data is available. Numbers of Curlews on the saltmarsh in particular are almost impossible to estimate with any confidence. In spite of the bird's size even quite large flocks can lurk in the vegetation, invisible unless flushed by a raptor, a low-flying aircraft or some other random occurrence. It is likely that at least 500 Curlews are present on the saltmarsh and shore on quite a few occasions between August and March and that numbers during this period seldom fall below 300.

On the fresh marshes Curlews begin to gather in late July. Numbers frequently reach 75 in August if the ground is not too dry and there are usually between 120 and 200 at the site from late September to the end of January. Numbers typically increase in February and March, presumably swelled by groups of migrants returning to upland breeding sites: 420 in March 1993 and 370 in February 1995 are the highest totals in recent years. Curlews largely disappear from Marshside by the end of April although occasional stragglers are seen in high tide roosts or on the fresh marsh in May and June.

SPOTTED REDSHANK *Tringa erythropus*

The Spotted Redshank is primarily an uncommon passage migrant at Marshside but there are also a few scattered records of birds in winter. The species can turn up almost anywhere, at Crossens channel, on the Sand Plant lagoons and adjacent creeks, or even on the open shore, but no more than two or three individuals are recorded in a typical year. Spring migrants are particularly rare; over the last 30 years there are about 15 widely dispersed records of single birds, or very occasionally small parties, in April and May; six together on M2 on 11 April 1998 is the largest group recorded.

Autumn Spotted Redshanks, a few of them adults still in full or partial summer plumage, but the majority smoky-grey juveniles are seen from early July to the beginning of October. Five in September 1980 and four in September 1991 are the largest groups recorded. In some years Spotted Redshanks in winter plumage are seen at Marshside on a few dates between mid-November and February and it seems likely that these are wintering birds wandering around the Ribble estuary. One in 1985-86 and two in 1991-92 spent the winter almost exclusively at Marshside and Crossens.

REDSHANK *Tringa totanus*

Redshanks are present all year round at Marshside as breeders, winter residents and passage migrants. They are one of our most conspicuous and noisy birds, so familiar as often to be taken for granted. Indeed, in contrast to those other common middle-sized waders the Lapwing and the Golden Plover, there is surprisingly little hard data available for Redshanks at Marshside, although long-term studies of the species's breeding biology were conducted on Banks marsh in the 1970s and 1980s. Unlike the species just mentioned Redshanks are normally widely distributed across both fresh- and saltmarshes as well as on the shore, a fact which makes counting difficult. Largest numbers are usually seen at high tide when substantial roosts may build up on both sides of the Marine Drive.

Redshanks at Marshside appear to be most numerous in mid-winter: 1,175 were present in high tide roosts on 21 December 1980, up to 1,000 on several dates in December 1982, and 1,170 on 5 January 1991. There are fewer counts at other times of year but 600 on 25 September 1983 and 850 on 12 April 1992 are typical of totals during the main passage periods. In recent years up to 38 pairs of Redshanks have nested on the fresh marsh. Numbers breeding on the saltmarsh are more difficult to assess but at least 15 birds were engaged in display flights in 1994; this had fallen to four by 1998. There is, unfortunately, a lack of year-by-year data on breeding numbers in the past but there is a widespread opinion among the local birding regulars that a significant decline has taken place, particularly on the saltings.

MARSH SANDPIPER *Tringa stagnatilis*

In an extraordinary series of occurrences, two juveniles arrived together on M2 on 5 August 1999; what was almost certainly a different juvenile was present on the following day. Two birds were seen again briefly on 7 August and one on 9 August. It remains unclear at time of writing how many individuals were involved, but this most elegant visitor from Central Asian breeding grounds was a very welcome, if totally unexpected, addition to the Marshside list.

GREENSHANK *Tringa nebularia*

The Greenshank is a passage migrant at Marshside in both spring and autumn and numbers recorded vary considerably from one year to the next. This fine upstanding Tringa with its ringing and evocative call is popular among the birding regulars. As with most of the scarcer northern breeding waders that pass through Britain and Ireland, Greenshanks are normally much more common in autumn than in spring. Like the Spotted Redshank they are catholic in choice of habitat at Marshside and are seen on both salt- and fresh marshes. Greenshanks in spring are normally recorded singly, mainly on the fresh marsh in late April and the first half of May; there were occurrences in 1978, 1979, 1983, 1987, in 1993-95 and in 1997-99.

In autumn Greenshanks are most often seen on the shore, by Crossens channel, or in the vicinity of the lagoons; records extend from early July to mid-November with the great majority in August and September. There are records in nearly every year since the mid-1970s, at least, and some substantial totals and even fair-sized flocks have been seen: 10 were on Crossens channel and saltmarsh on 30 August 1980 and 21 there on 9 September 1981, 27 were counted on 3 September 1988 and there was a total of 17 in September 1991. Since then eight birds on Crossens channel on 2 September 1996 is the highest number recorded. A single bird on 19 January 1988, and one by Crossens channel on several dates in late December 1998 are the only mid-winter occurrences in the last two decades.

LESSER YELLOWLEGS *Tringa flavipes*

One which frequented the Sand Plant lagoons and adjacent creeks on many dates between 18 October and 2 December 1997 was our first record of this graceful Nearctic vagrant. This was presumably one of the two birds that were present intermittently on nearby Banks marsh from late December 1997 through 1998 and into 1999; one or other of these visited Marshside again on nine occasions during 1998, in January, March, April, May, July and November.

GREEN SANDPIPER *Tringa ochropus*

Until very recently the Green Sandpiper was a scarce passage migrant at Marshside with only marginally more records in the previous two decades than the Wood Sandpiper, which is by far the rarer of the two species in Lancashire as a whole. From 1978 to 1997 there were only three records in spring involving five birds, all between 8 and 27 April, and 11 records in autumn involving 12 birds, between 14 July and 15 September. In 1998 and 1999 the number of autumn records in particular increased dramatically, probably due to the presence of the Sandgrounders Hide between the M2 creeks and the lagoons, this species's favourite haunt; scraping and ditching work on the marsh must also have improved the site's appeal for this species. Two birds were recorded in spring 1999, and at least a dozen in autumn. One on M1 on 31 December 1998 is our only mid-winter record.

WOOD SANDPIPER *Tringa glareola*

Until about 1980 the Wood Sandpiper was a scarce but fairly regular migrant, especially in autumn; in 1976, for example, up to three birds were on Crossens channel in late August-early September while another spent several days at rain pools on Hesketh Golf Course in early November. Numbers in spring were typically lower but two were on M2 from 10 to 13 May 1979. During the 1980s the frequency of records in autumn decreased with one in August 1980 and two in July-August in 1986 and 1988. There was only one spring occurrence, of two birds on M1 between 7 and 10 May 1989.

The decline seems to be levelling out, however; there have been seven records of Wood Sandpiper in the 1990s, single birds on 10 September 1991, 11 May 1993, 18 May 1994, 21 June 1995, 5 July 1996 and on 23 August and 17-18 September 1997.

COMMON SANDPIPER *Actitis hypoleucos*

At Marshside the Common Sandpiper is an uncommon though regular migrant in spring and autumn and its tendency to favour the margins of the Sand Plant lagoons and the scrape in front of the Hide makes the species familiar even to casual birders. Numbers vary a great deal from year to year but there is some evidence of an overall decrease over the last quarter century. In spring Common Sandpipers normally pass through between the middle of April and about 20 May. Estimates of numbers are made difficult by the probable presence of long-staying individuals in most seasons but 7-9 birds were recorded on average in the early 1980s and 4-6 in the early 1990s. A notable increase to some 15 spring migrants took place in 1995 and this was maintained in spring 1996, when at least 16 were recorded, but numbers slumped again in 1997 and 1998. Spring 1999, with 18 migrants recorded, was the best of the decade, however. One or two birds may have summered on M2 in 1995.

Autumn Common Sandpipers are found in a wider range of habitats than those in spring; in addition to the Sand Plant lagoons and adjacent creeks passage migrants are seen at Crossens channel and occasionally in tidal pools on the shore by the sand road. Occurrences extend from the end of June to mid-October with most recorded in July and August, and an overall decrease in seasonal totals is evident since the mid-1980s. In the early and mid-1990s around 5-6 Common Sandpipers were seen at Marshside each autumn and multiple occurrences were rare; this was a mere shadow of the 16 recorded on 15 August 1977, the 12 on 16 August 1981, or even of the seven on 14 July 1982. A minor upsurge in totals for autumn 1996 (22) and 1997 (40) was not sustained in 1998, but a total of 60 in autumn 1999 with up to six birds present on several dates is very encouraging. A single bird was present at Crossens Water Treatment Plant from 17 December 1980 to mid-March 1981; this is the only instance of overwintering during the last 20 years.

TURNSTONE *Arenaria interpres*

Turnstones are present at Marshside in small numbers throughout the year and may be seen feeding by the sand road as well as in high tide roosts on the shore south of the Sand Plant. Occasionally, when very high tides and onshore gales coincide a few roost on the grazed portion of Crossens saltmarsh or even on the fresh marshes. The Turnstone is a passage migrant as well as a winter resident; there are relatively few counts specific to Marshside but 37 were present on 14 June 1980, 43 on 25 November 1981, 83 on 2 January 1983 and 230 on 25 July 1994. There is strong evidence of a decline both in winter and in passage numbers since the mid-1990s.

RED-NECKED PHALAROPE *Phalaropus lobatus*

A fresh corpse picked up on M2 on 21 October 1978 is the only documented record.

GREY PHALAROPE *Phalaropus fulicarius*

There are five recent records at Marshside, four involving storm-driven autumn migrants. One was present on 29 November 1968, a juvenile and a first-winter were on the M1 floods on 30 September and 1 October 1978 (the juvenile bird was found dead on 4 October), one was seen briefly on 9 October 1980, and another frequented the muddy shore immediately south of the Sand Plant from 4 October until early November 1981. A first-winter bird turned up in calm weather on 8 December 1979 and stayed in the Sand Plant area until 31 December.

POMARINE SKUA *Stercorarius pomarinus*

Pomarine Skuas are regularly recorded on autumn passage off Formby Point with between three and six birds seen annually in recent years. There have been seven records of this powerful piratical seabird at Marshside during the last quarter century involving a total of 13 birds. Most were autumn migrants driven into the mouth of the Ribble by onshore winds between late August and early November, including a flock of six off the Sand Plant peninsula on 17 October 1991. There are two recent occurrences of Pomarine Skuas at Marshside in winter, one on 23 November 1986 and a juvenile bird which was present intermittently from 30 December 1994 to 4 January 1995, occasionally pursuing gulls and waders over the shore and the fresh marshes.

ARCTIC SKUA *Stercorarius parasiticus*

Although far more Arctic Skuas than Pomarines are normally seen during seawatches on the Lancashire coast, totals of the two species recorded at Marshside are roughly equal. Arctic Skuas are scarce and irregular visitors mainly in autumn. Most are seen passing along the shore or over the far edge of the saltmarsh at high tide and they rarely linger, although an immature bird spent over three weeks on Crossens Outer in November-December 1977. Apart from that individual and another on 12 March 1985 all the 12 birds since 1976 have been seen between mid-August and mid-October. Three on 21 August 1988 is the largest day total.

GREAT SKUA *Cartharacta skua*

Like its two smaller relatives the Great Skua is a very scarce visitor to Marshside but records of the present species are somewhat less restricted to the main autumn passage period. Of 12 or 13 Great Skuas recorded since 1970 seven were seen between 28 July and 8 November, but one spent several weeks on Crossens saltmarsh in spring 1972 and four or five individuals have been seen in mid-winter: on 10 December 1977, on 15 January 1983 and possibly the same bird on 2 February, on 12 January 1990, and on 1 January 1995. All sightings have been over the estuary or the saltmarsh, usually at high tide and during onshore winds.

MEDITERRANEAN GULL *Larus melanocephalus*

Unknown in Lancashire before 1968 the Mediterranean Gull was already averaging three records a year by the early 1970s (Spencer, 1973). This handsome gull is now regularly seen at several sites in Lancashire and N.Merseyside and is virtually resident in small numbers at Seaforth NR. At Marshside the first Mediterranean Gull, an adult in winter plumage, was recorded on 23 November 1975 and there were seven further sightings of individual birds up to 1989, distributed across all four seasons. There has been a distinct quickening in the rate of occurrences at Marshside during the 1990s with one in 1992, six in 1993, two in 1994, two in 1995, five in 1996, ten in 1997 (seven on the same day, 25 February), three in 1998, including several sightings of an adult bird in late May and early June, and one in 1999, also in spring.

The great majority of records are in winter but four birds have been seen in April, one in May/June and one in August. Mediterranean Gulls at Marshside are almost invariably found in high-tide roosts of Black-headed or Common Gulls, either on the fresh marshes or, more rarely, on the shore; most visits are of brief duration. Of the 30 records for which age data is available 26 were adults, three second-years, and one first-year.

LITTLE GULL *Larus minutus*

Before 1965 the Little Gull was a scarce visitor to Lancashire with an average of only three records a year during 1954-64 (Spencer,1973). Numbers increased dramatically after the mid-1960s and the species is now a widespread passage migrant in spring and a fairly frequent visitor at other times, mainly at coastal sites. Seaforth NR at the mouth of the Mersey is the regional, indeed often the national, mecca for Little Gulls; up to 200 are commonly recorded there in spring but smaller numbers are frequently present at other seasons. At Marshside Little Gulls are uncommon and sporadic visitors although small flocks are occasionally recorded; occurrences in the last quarter century fall into two distinct seasonal clusters.

Primarily the species is an irregular transient on spring passage with records in 1977, 1980, 1981, 1989, 1991, 1995, 1996 and 1998. Birds are usually seen hawking over the M1 floods or the Sand Plant lagoons and they may linger for several days or even longer before moving on. Multiple occurrences are not untypical: seven were present on 15 April 1980, eight on 4 April 1981, four on 26 March 1989 and five in late May-early June 1995. Most of the visitors in early spring have been adult birds but the late migrants in 1995 and 1998 were all first-summer individuals.

Apart from isolated occurrences of eight on 4 July 1965 and seven on 11 September 1992, all the Little Gulls recorded at Marshside outside the spring migration period have been storm-battered refugees in winter, seeking shelter in the massive gull roosts that assemble at such times on the inland side of the Marine Drive: ten were seen on 17 December 1979, another ten on 1 February 1982, up to 20 on 1 February 1983, four in particularly wild conditions on 26-27 February 1990, seven on 11 September 1992, a total of five in late January 1993, up to eight during the first week of January 1998, and three on 4 January 1999.

BLACK-HEADED GULL *Larus ridibundus*

Black-headed Gulls are abundant at Marshside at all times of the year. Several thousands of pairs breed, or attempt to breed, annually on Banks and Hesketh out-marshes and birds from these colonies are common visitors during the spring and summer months, with juveniles accompanying their parents in July and August. In 1998 there were up to ten nests for the first time on Crossens saltmarsh and at least one pair nested on M2 in the vicinity of Polly's Creek; two or three young were reared. Two pairs nested on M2 in 1999, at least 30 on the adjacent saltmarsh, and an unknown but probably large number on the outer reaches of Crossens; it is doubtful if many clutches survived the spring tides, however.

Combinations of high tides and onshore gales in winter produce immense roosts of gulls both on the shore and on the fresh marshes, and Black-headed Gulls usually constitute a substantial proportion of these. 6,500 on 24 January 1993, 12,500 on 19 December 1993, 4,000 on 30 January 1994 and 13,900 on 25 February 1997 are typical of winter roost counts in recent years. In less spectacular conditions there are, nevertheless, always good numbers of Black-headed Gulls about, no doubt including many non-breeders and passage migrants depending on the season. Leucistic and other aberrant individuals are occasionally encountered.

RING-BILLED GULL *Larus delawarensis*

There have been nearly 80 records across Lancashire of this wanderer from North America since its first occurrence at the Alt estuary on 13 April 1979. Only four of these have been at Marshside, a second-summer bird on 13 August 1986, one, age unspecified on 16-17 January 1993 and second-winter birds on 12 November 1996 and 2 February 1998.

COMMON GULL *Larus canus*

The Common Gull is the third most numerous gull species at Marshside after Herring and Black-headed Gulls. It is a numerous winter visitor and a migrant in both spring and autumn. As with most of our gull species numbers are usually at their highest in winter, when hundreds may be seen in high tide roosts. 4,700 on 18 February 1996 is the largest count in recent years but totals of 1,500 to 2,000 are not unusual in boisterous weather between December and March. An adult found freshly dead on the Marine Drive on 9 December 1998 had been ringed as a nestling in Vaasa, Finland on 6 July 1996. The species is relatively scarce in summer and early autumn although even then a few loafers are nearly always present on the shore or by the Sand Plant lagoons.

LESSER BLACK-BACKED GULL *Larus fuscus*

In contrast to the other large gull species Lesser Black-backs at Marshside are usually at least as numerous between April and September as they are in winter, although most of our count data are from the big high tide gull roosts that are such an exciting feature of the December-March period. There are typically around 50 to 80 Lesser Black-backs in these gatherings but in recent winters an exceptional 360 were present on 30 January 1994 and 230 on 24 January 1995. A few individuals of the Scandinavian subspecies fuscus have been identified in the winter roosts; four on 23 December 1991 is the most recent occurrence.

From late spring to mid-autumn Lesser Black-backs are ever-present at Marshside, sailing apparently aimlessly overhead or loafing on the fresh marsh and on the shore. Counts are lacking but 30 to 40 may be seen on almost any visit and the great majority are adult birds. Up to 3,000 pairs normally nest on Banks and Hesketh saltmarshes and this population presumably accounts for the majority of visitors to Marshside, not to mention the squadrons permanently cruising over Southport town itself or moving off Formby Point throughout the summer months.

YELLOW-LEGGED GULL *Larus (cachinnans) michahellis*

There have been five records of this close relative of the Lesser Black-backed and Herring Gulls, single adults on 12-16 February 1990, 17 September and 10 December 1993, one, age unknown on 29 July 1996, and an adult and a first-winter in a roost on M1 on 3 March 1998.

HERRING GULL *Larus argentatus*

Although the Herring Gull is frequently outnumbered by other gulls at Marshside, peak counts of this species in mid-winter roosts far exceed those of any of the others. In recent years 18,000 were present on 22 December 1991, 12,000 on 24 January 1993, 23,000 on 19 December 1993, 15,000 on 30 January 1994, 14,200 on 24 January 1995 and 13,500 on 25 February 1997. Small numbers of the nominate Scandinavian race are quite regularly identified in these gatherings; ones and twos are the norm but up to ten were seen on 15 January 1993 and 12 on 10 December of the same year. Herring Gulls may be seen at Marshside throughout the year but in contrast to Lesser Black-backs they tend to be least common in mid-summer, in spite of the fact that over 1,000 pairs normally nest on Banks and Hesketh out-marshes.

ICELAND GULL *Larus glaucoides*

Until quite recently the Iceland Gull was an exceedingly rare visitor to Marshside with only one documented record, a second-year bird on 29 April 1978. There were, however, two in 1994, a first-year on 13 March and a second-summer on 3 April; three were recorded in 1995, an adult and a second-winter together on 1 January and a third-summer on 14 April. One was on M1 on 6 November 1996 and a second-winter spent half an hour in front of the Hide on 18 September 1997; a third-winter was on M2 on 31 January 1998, and three were recorded in February and March 1999, a first-winter, a second-winter and an adult. Our first Kumlien's Gull L. g. kumlieni from Arctic Canada was on M1 on 5 March 1997. It is perhaps too early to reclassify the Iceland Gull definitively as a regular winter and passage visitor, but a pattern seems to be becoming established. All the birds since 1994 have been seen in gull roosts on the inland side of the Marine Drive embankment.

GLAUCOUS GULL *Larus hyperboreus*

The always impressive and sought-after Glaucous Gull is a scarce though fairly regular visitor to Marshside; records have been annual since 1989. The great majority are seen in winter or early spring gull roosts but there have been occurrences in all months except July. Numbers vary greatly from year to year; two or three birds is the norm but five were recorded in 1993, four of them in January, 13 were seen in 1994 with an unprecedented seven different individuals during the second half of January, and there were visits by five birds in 1998, in March, April and November.

Most visits by Glaucous Gulls to Marshside are brief, of a day's duration or less, but several have stayed for lengthy periods including an exceptionally large adult bird present, mainly on M2, from late December 1978 to mid-February 1979. Of the 46 documented records since 1977 14 have been first-years, six second-years, four third-years, one fourth-year, and the remaining 21 adult birds.

GREAT BLACK-BACKED GULL *Larus marinus*

Although the Great Black-back is normally the least numerous of the five common gulls at Marshside it is, nevertheless, a familiar species and may be encountered in small numbers on the shore or on the fresh marsh at any time of year. Peak counts in roosts in recent years include 280 on 17 December 1993, 155 on 24 January 1994, 315 on 24 January 1995 and 440 on 20 August 1997.

ROSS'S GULL *Rhodostethia rosea*

A winter adult was on the tideline briefly on 16 January 1983 before moving off southwards.

KITTIWAKE *Rissa tridactyla*

Although Kittiwakes are fairly regular off Formby Point between early April and the end of October there are fewer documented records at Marshside of this familiar coastal gull than there are of either Mediterranean or Glaucous Gulls. Although the species is probably under-recorded both on passage along the shore and in storm-driven winter roosts, Kittiwakes are undoubtedly no more than scarce and sporadic visitors to this site. A flock of 45 after violent north-westerly gales on 1 February and one offshore on 1 April 1983, three on 11 September 1992, three in a large high tide roost on 23 January 1993, up to four on several dates in the first week of January 1998, one on M2 on 16 December 1999 and 15 over the tide in a gale on Christmas Day 1999 are the only recent records.

SANDWICH TERN *Sterna sandvicensis*

The Sandwich Tern is an uncommon and irregular visitor to Marshside on spring and autumn migration. Sightings are virtually confined to the shoreline and the numbers recorded vary from year to year. This is normally the earliest of the terns to arrive in our waters in spring and Marshside records include three on 22 April 1978 and two on 12 April 1979. Most Sandwich Terns are seen on passage in early autumn, but the small high-tide tern roosts that used to gather on the foreshore as recently as 1994 have all but vanished; 12 birds in August/September 1999 was the best showing for several years.

ROSEATE TERN *Sterna dougallii*

There are only three documented records of Roseate Tern at Marshside, single birds on 20 May 1975 and 27 August 1980 and three on 31 August 1981. This species has been in decline on the coasts of North-West Europe as a whole in recent decades but the establishment of a thriving colony in Dublin Bay since the late 1980s has been followed by several sightings during the 1990s at Seaforth NR and off Formby Point. The virtual disappearance of autumn tern roosts from the shore at Marshside, however, makes the likelihood of future records rather remote.

COMMON TERN *Sterna hirundo*

Though possibly declining, this is the most numerous tern species at Marshside as it is all along the Lancashire coast. A nesting colony established on Banks and Hesketh saltmarshes in the late 1960s fluctuates in size from year to year, from a few score to several hundred pairs. A small number of Common Terns, presumably associated with the Banks colony, may be seen regularly around the Sand Plant lagoons and adjacent creeks from May to July.

Up to the mid-1990s numbers on autumn passage at Marshside did not typically reach double figures until late July; from then until mid-September individuals and small flocks were frequently observed moving over the estuary or roosting with gulls and waders on the beach at high tide. Numbers varied considerably from year to year: up to 450 were present on several dates in early September 1979 and at least 1,100 were counted on 28 August 1984. Totals in the early 1990s were appreciably smaller, with 220 on 19 August 1993 and 80 on 25 July 1994 the highest numbers recorded; since 1995 very few have been seen in any year, and a total of 30 between July and September 1999 was exceptional by the reduced standards of the late 1990s.

ARCTIC TERN *Sterna paradisaea*

Arctic Terns are scarce visitors to Marshside on spring and autumn passage, and several years may go by without a record. Spring occurrences usually involve fast-moving individuals or small parties on migration; several have been seen over the fresh marsh like the group of three that raced through on 14 May 1995. Autumn sightings are normally on the shore or over the estuary at high tide; three on 21 August 1989, three on 12 September 1991, four on 19 August 1993 and six on 28 July and 12 on 23 September 1999 are the highest numbers recorded in recent years.

LITTLE TERN *Sterna albifrons*

Little Terns are scarce visitors to Marshside on migration in both spring and autumn. Birds are almost invariably seen moving immediately offshore or in high tide roosts. Spring migrants occur in about one season in four and records are confined to the period from late April to mid-May; five together on the shore south of the Sand Plant on 24 April 1994 is the highest number recorded in recent decades. In autumn Little Terns have become very scarce since the early 1990s; they typically occur in late July and August and eight on 7 August 1994 is the largest day-total.

BLACK TERN *Chlidonias niger*

This attractive and sought-after migrant is a scarce and irregular visitor to Marshside. Since 1976 only 13 have been seen in spring (of which seven occurred in 1994) and 12 in autumn. Spring birds, usually in immaculate breeding plumage, are seen over the remnant floods and creeks on the fresh marsh. Visits are occasionally prolonged as when single birds remained for four days in late April 1983 and for two days in early May 1994. Occurrences in autumn extend from early July to early October and birds are usually seen only briefly, either hawking insects over the saltmarsh and tidal creeks or moving along the tideline. Adults and juveniles have been recorded with roughly equal frequency.

WHITE-WINGED BLACK TERN *Chlidonias leucopterus*

An adult was seen over Crossens on 23 August 1968 during a minor influx into North-West England of this rare marsh tern. This is the only record.

GUILLEMOT *Uria aalge*

Although the Guillemot is normally much the commonest auk species on the coast of Lancashire there are only two recent records for Marshside. An immature bird was found dying on the Sand Plant lagoons on 25 September 1987 and one was offshore on the tide on 1 March 1990. As in the case of other seabird species it is likely that Guillemots are occasionally overlooked on the estuary, particularly during rough weather.

BLACK GUILLEMOT *Cepphus grylle*

One during a gale on 3 September 1955 is the only Marshside record of this very scarce visitor to the South-West Lancashire coast.

LITTLE AUK *Alle alle*

The corpse of a Little Auk, long-dead, was found on Crossens saltmarsh on 29 April 1984. It has been suggested that this was the same individual released at Marshside on 10 February of that year, having been rehabilitated at Martin Mere WWT reserve after being brought there exhausted following January gales. This cannot be verified, however, and the released bird was reported as flying off strongly out to sea following its liberation.

FERAL PIGEON *Columba livia*

Three or four pairs regularly nested in the old Sand Plant tower prior to its demolition in 1994; the situation since then is unclear. A few Feral Pigeons are seen quite frequently on the fresh marshes and around the lagoons but little attention is paid to them by birders and no estimates of numbers are available. The picture is complicated by the fact that racing pigeons from nearby lofts are also regular visitors to the area.

STOCK DOVE *Columba oenas*

The story of this unobtrusive but elegant pigeon at Marshside over recent years is one of quite sudden and unexplained decline. Before about 1986 Stock Doves were numerous visitors in winter, mainly to the saltmarshes. Flocks of over 50 were a common sight and an increase appeared to have taken place as recently as the early 1980s. Between 1976 and 1983 peak counts of 150-200 were typical of the period from mid-November to late February with 350 on Crossens saltmarsh on 20 February 1980 the highest number recorded; by 1984 and

1985 January and February totals were occasionally reaching 400. By the 1987-88 winter, however, the species was already in full retreat: 65 birds in late January 1988 was the highest count that winter and the peak for the next year was a mere 35 Stock Doves on the saltmarsh on 12 February 1989.

Since the early 1990s it is clear that numbers of Stock Doves at Marshside in winter have fallen to an extremely low level; 22 on 17 February 1994 is the largest count recorded since 1991. Both locally in the South-West Lancashire area and more widely throughout Britain and Ireland a marked decrease in numbers has been noted (Gibbons, Reid & Chapman, 1993). As well as our diminished winter population a few birds are still occasionally seen at other seasons, feeding on the fresh marshes or on Hesketh Golf Course. Breeding has never been recorded.

WOODPIGEON *Columbus palumbus*

Although they are seldom counted Woodpigeons are common at Marshside, particularly on Hesketh Golf Course and on the fresh marshes. Large loose flocks, occasionally numbering over 50 birds, may be seen feeding at almost any time of year although numbers tend to be highest in spring and early summer and lowest in mid-winter. Up to 15 pairs have nested on the Golf Course each year during the 1990s, and since 1996 up to three pairs have bred annually in bushes on the Marine Drive embankment.

COLLARED DOVE *Streptopelia decaocto*

The Collared Dove was first recorded, and bred, in Lancashire in 1961 and by 1965 the species was already established in pockets throughout the Southport area. Numbers in the vicinity of Marshside appear to have stabilized by the early 1980s. The species is an occasional visitor to the fresh marshes and Hesketh Golf Course from nearby suburban gardens and from the Crossens Water Treatment Plant compound, where a few pairs occasionally breed. In recent years parties of up to 15 birds have been seen frequently in late summer feeding on the banks of Crossens channel immediately opposite the Water Treatment Plant.

TURTLE DOVE *Streptopelia turtur*

The Turtle Dove established a breeding foothold in Lancashire south of the Ribble during the early years of the twentieth century (Oakes,1953). This was maintained and consolidated into the 1950s and 1960s, but numbers both nesting and on passage have declined dramatically since the early 1980s. The Turtle Dove appears always to have been a scarce and irregular migrant at Marshside. Almost all the recent spring records have been in May with birds seen on the Golf Course, the M1 embankment, and the Sand Plant peninsula. Two on 11 May 1993 and on 9 May 1998, and two or three on M2 on 13 May 1999 are the only multiple occurrences. In fact, in most springs since 1980 there have been no records at all at Marshside, and two birds on 11 August 1993 and singles on 18 August 1996 and on 19 August 1998 are the only recent autumn sightings.

RING-NECKED PARAKEET *Psittacula krameri*

The first Marshside records of this range-expanding feral resident were of single birds on Hesketh Golf Course on 19 February and 29 April 1998.

CUCKOO *Cuculus canorus*

At Marshside the Cuckoo is a migrant, predominantly in spring, whose numbers tend to fluctuate quite widely from year to year. There is clear evidence of an overall decrease in numbers since the early 1980s. Most Cuckoos are seen, or more often heard, on Hesketh Golf Course but an occasional migrant may be encountered just about anywhere. Recent spring records extend from 22 April to 30 May with most in early- to mid-May. No more than three or four Cuckoos are recorded nowadays even in a good spring; autumn migrants are even scarcer with only four or five juvenile birds seen, all in August, over the past 15 years. There is no evidence of breeding.

BARN OWL *Tyto alba*

A scarce and irregular visitor which has been recorded at all times of the year. The species remains reasonably numerous on the nearby mosslands and may well be under-recorded at Marshside due to its nocturnal habits, but for what the information is worth there have been only about 20 occurrences, almost all of single birds, since 1980. In December 1999 a bird took to roosting on several days in the Sand Plant buildings, prompting hopes that the status of this atmospheric night hunter at Marshside may be about to change.

LITTLE OWL *Athene noctua*

Although the Little Owl is resident in fair numbers on the South-West Lancashire mosslands the species is rare at Marshside. Occasionally a road-kill corpse is picked up along the Marine Drive; four of these were found during 1979 alone, but there have been very few since. The most recent records of a live Little Owl are of one seen on several dates between 22 October and mid-November 1992 in various parts of both fresh and saltmarshes, and singles on the Golf Course from 22 to 24 December 1996 and on 12 September 1999.

TAWNY OWL *Strix aluco*

Tawny Owls are fairly common residents throughout the Southport area, particularly in districts with well-wooded parks and large gardens. There are several such immediately adjacent to Marshside and it is likely that the species is a regular though unobserved nocturnal visitor, especially to Hesketh Golf Course. The only definite records in recent years are of birds calling at dusk on the Golf Course in late July and August in 1994 and 1995, which are likely to have been recently-fledged youngsters dispersing from nearby nest sites, one seen on several mornings in mid-January 1998 on the goal-posts on Stanley School field, and one on the Golf Course early on 5 September 1999.

SHORT-EARED OWL *Asio flammeus*

As recently as the late 1940s Short-eared Owls were considered to be no more than casual visitors in very small numbers to the Ribble estuary in winter (Oakes, 1953). By the early 1970s, at least, the species was a regular winter resident and passage migrant in some numbers, mainly on the south side of the estuary from the Douglas River estuary to Crossens. In winter the Owls frequented both the saltmarshes and the reclaimed farmland areas and were often to be seen beating low over the ground on bleak afternoons in search of rodents and small birds. At Marshside-Crossens in particular Short-eared Owls favoured the saltmarsh from which numbers were regularly flushed by high tides, wildfowlers and birders. Eight to ten birds were normally present in mid-winter in the late 1970s with at least 12 seen together in December 1978. Winter visitors usually arrived in November and departed in February and March, but individuals on passage were regularly seen in late April and May.

This situation remained largely unchanged in the early 1980s and a total of 17 in November 1982 was the highest-ever count of Short-eared Owls at Marshside. By the middle of the decade, however, numbers had declined dramatically. Fair-sized winter roosts were occasionally still recorded on the Inner Ribble but a general decrease was noted in the Lancashire Bird Report for 1987 (Jones, 1988). From 1984 to 1993 no more than five Short-eared Owls were seen on any date at Marshside. The great majority of records were of single birds in mid-winter and occurrences during the late spring passage period dwindled virtually to nothing.

Since the early 1990s there has been something of a recovery as far as peak counts are concerned: eight were seen in January 1993 and up to 10 birds were flushed by exceptionally high tides in January and in December 1995; there were nine over the saltmarsh at high tide on 9 February 1997. In contrast to the situation in the past, however, the Owls, if present, are very seldom seen nowadays except during unusually powerful tidal surges; regular patrolling of the saltmarsh seems to have been abandoned. How the birds obtain their prey is, therefore, something of a mystery; perhaps a more passive style of hunting has been adopted or it may be the case that the birds have abandoned diurnal activities in favour of an exclusively nocturnal strategy.

NIGHTJAR *Caprimulgus europaeus*

The Nightjar has decreased catastrophically in numbers over North-West England as a whole since the 1960s. There are only three recent records at Marshside, of single migrants over the Sand Plant lagoons on 23 May 1976 and 26 September 1981, and a bird heard briefly churring on the Golf Course at dawn on 16 May 1999.

SWIFT *Apus apus*

The Swift is a common, in some years an abundant, spring migrant at Marshside and birds from nearby urban breeding sites are present over the area in varying numbers throughout the summer. The first arrivals are usually seen in the last few days of April with one on 22 April 1996 the earliest in recent decades. Passage continues throughout the whole of May and in some years, when a strong southerly or south-easterly airstream becomes established particularly in the first half of the month, very large movements of Swifts are recorded. Several hundred per hour passed on 11 and 12 May 1980, 1,250 were recorded in 4.5 hours on 11 May and 1,200 in 3 hours on 24 May 1993, and 1,800 moved north in 4.5 hours on 11 May 1995.

During periods of settled weather in midsummer substantial feeding flocks of Swifts occasionally build up over the saltmarsh and over 700 were counted on 26 June 1992. Autumn migration usually passes almost unnoticed at Marshside. A few, usually small, groups of obvious migrants are seen in the first half of August but very few Swifts are recorded after the beginning of September.

KINGFISHER *Alcedo atthis*

Kingfishers are scarce and irregular visitors mainly to the Sand Plant lagoons and adjacent creeks, although birds are also occasionally seen perched by tidal channels on the saltmarsh. In the past quarter century Kingfishers have been recorded in March and April as well as on a range of dates between early August and late January. Most visits are single day affairs but a few birds have remained for quite lengthy periods, usually in autumn and winter. In spite of the Kingfisher's brilliant plumage birds at Marshside are sometimes remarkably elusive, and often only the shrill call is heard as the bird itself flies unseen along the creeks, screened by vegetation.

WRYNECK *Jynx torquilla*

One calling on Hesketh Golf Course on 1 May 1981 is the only record.

GREAT SPOTTED WOODPECKER *Dendrocopos major*

During the late 1970s and 1980s there were occasional sightings of Great Spotted Woodpeckers on Hesketh Golf Course mainly in autumn and winter. In 1993 a pair bred successfully on the Golf Course in trees bordering Fleetwood Road and this was repeated in 1994 and, at different nest sites, from 1995 to 1997. In 1998-99 nesting took place just outside our recording area, on the inland side of Fleetwood Road, but frequent visits to the Golf Course have continued; birds are also occasionally recorded in the willows along the SSSI ditch. Individuals seen in bushes along the Marine Drive in September and October 1996 and in October 1997 probably originated from further afield.

SHORE LARK *Eremophila alpestris*

There are four or five records of this most attractive passerine, now a very rare visitor to Lancashire; a single bird was seen along the shoreline at both Marshside and Crossens on a number of dates in November and December 1973, another was on the foreshore near Hesketh Road on 9 and 12-13 December 1998, and there were two or three records in early 1999, three birds on the shore on 16 January and singles, or the same individual, by the lagoons and on M2 on 16 and 24 February.

SKYLARK *Alauda arvensis*

The Skylark is one of Marshside's most familiar passerines. It is a common passage migrant and winter visitor and the first song-flights of territorial males in mid-February are one of the earliest and most welcome signs of the approach of spring. Although the species is still fairly numerous and likely to be seen on any visit at any time of year, Skylark numbers at Marshside have declined very significantly over the past two decades, in particular since the mid-1980s. This decrease has affected wintering and passage totals as well as the nesting population. During the 1976-77 and 1978-79 winters, for example, literally vast numbers were to be seen on and over the saltmarsh. Counting was, unfortunately, rarely even attempted but many thousands must have been present and the spectacle excited little comment among the birding regulars at the time. Numbers seem to have fallen somewhat in the early 1980s but up to 6,000 were estimated to be on the saltmarsh as recently as January and February 1986. Since the end of the 1980s few winter counts have exceeded 800, and totals of 250-400 weremuch more typical of the situation in the late 1990s. There are very few published counts of migrant Skylarks extant from the 1976-86 period but even a cursory examination of field notes indicates that very large movements were quite common in March and again in late September and October, with over 1,000 birds in a day fairly normal in autumn. Passage movements nowadays very rarely exceed 200 in a day in March or 600 in October.

There is a lack of published data on breeding numbers at Marshside over the same period but regulars retain vivid if unquantifiable memories of an all-pervasive chorus of song on still days in early spring. About 86 pairs were counted on M1/M2/Crossens Inner in April 1983 with perhaps another 35 pairs on the saltmarsh. By 1997-1998 there were probably no more than 40 pairs in the entire area, and the 36 pairs in 1999 was the lowest number yet recorded.

SAND MARTIN *Riparia riparia*

The Sand Martin is the scarcest but also normally the first to arrive of the three familiar hirundines at Marshside. One on 14 March 1999 is the earliest record in the last quarter century but there were late March birds in 1977, 1982, 1987, 1993, 1995, 1996 and 1998. Most Sand Martins pass through in April but an exceptional 100-plus were seen on 11 May 1993, a day of spectacular migration by many passerine species, and occasional birds are recorded up to the first week of June.

Both timing and numbers of Sand Martins involved in spring passage vary enormously from one year to the next. Prevailing weather conditions appear to be the main determinant, with migration late and sparse in cold, wet seasons and many more birds seen when warm southerly or south-easterly airstreams are dominant. In the generally benign spring of 1995, for example, movements extended from 23 March to 19 May and at least 50 birds were recorded with a peak of 14 on 6 April; 95 passed in spring 1996, with a peak of 23 on 24 April, and there were 74 in 1999.

Autumn movements of Sand Martins at Marshside are normally light although it must be admitted that observer coverage is usually much less intensive in July and August than during the exciting spring days. In most years up to 30 birds are seen between mid-July and late August, but over 390 were recorded in the exceptional autumn of 1996, with a peak of 260 on 29 July. One on 23 September 1999 was the latest seen in recent years.

SWALLOW *Hirundo rustica*

The Swallow is the commonest hirundine at Marshside and in favourable weather some spectacular movements are seen in May and, more rarely, in late August and early September. As with other aerial feeders the timing and volume of spring passage is greatly dependent on the prevailing weather patterns. In a typical year the first very welcome migrants are seen in the first week of April, but late March birds have become more frequent in the late 1990s; one on 15 March 1999 is the earliest recorded arrival. Movements continue intermittently until the first week of June with peak numbers normally between about 25 April and 20 May. 600 in 3 hours on 13 May 1980, 700 in 4 hours on 11 May 1982 and 650 in 4 hours on 11 May 1993 are the highest counts made during the last quarter century. Swallows nest fairly commonly on the mosses and in smaller numbers in Churchtown and other residential areas and up to 20 birds may be seen over

Marshside on midsummer days. Since 1995 one or two pairs have bred annually in the Sand Plant buildings.

Autumn migration normally begins in early August and continues sporadically until late September. There are a few records of birds in October and November with one on 23 November 1997 the latest in recent decades. Autumn movements are usually less spectacular than those in spring but some noteworthy totals have been recorded in recent years. 100 to 150 are typical of high counts in late August and early September but 560 streaming south across the Ribble estuary in just 45 minutes on 22 August 1994 was exceptional; 3,000 on 16 September 1997 is easily the highest day total recorded at any season.

HOUSE MARTIN *Delichon urbica*

House Martins usually arrive later in spring than their two close relatives and in cold and dismal seasons, as in 1991, none may be seen until early in May. In more typical years the first House Martins are recorded in mid-April and passage continues in fits and starts until the end of May. Numbers in spring vary enormously from year to year and only occasionally approach those of the Swallow, but up to 50 may be seen on a good day and there are some exceptional peak counts including 300 on 12 May 1980, 250 on 24 May 1993, and 550 in 4.5 hours on 11 May 1995.

House Martins nest in fair numbers in residential areas immediately adjacent to Marshside and birds are continuously present over the area throughout the summer; counts of 40-50 birds are by no means unusual. Autumn passage usually gets under way in mid- to late August and reaches its peak in mid-September when over 200 birds in a day may occasionally be seen. There are a few recent records of late stragglers in October with two on 25 and one on 29 October 1995 the latest in the last two decades.

RICHARD'S PIPIT *Anthus novaeseelandiae*

There are two records of this large Asian pipit at Marshside: one was on the Sand Plant peninsula on 16 January 1995 and another fed with migrant Skylarks on the shore south of the Sand Plant on 1 November of the same year.

TREE PIPIT *Anthus trivialis*

The Tree Pipit is an uncommon though annual spring migrant at Marshside, much scarcer in autumn. Tree Pipits may be encountered almost anywhere but most are seen singly on Hesketh Golf Course, along the verges of the Marine Drive and on the Sand Plant peninsula. The earliest spring records are of singles on 28 March in 1997 and 1998, but most pass through between mid-April and mid-May; up to 20 may be recorded in a good season. Few linger for more than a matter of minutes and many are fly-over records identified only by the species's distinctive wheezing call. A singing male was present on the Golf Course for several weeks in May and June 1981 and another or the same bird again in May 1982, although there was no evidence of breeding. In most years there are one or two records of migrant Tree Pipits in autumn; almost all have been seen in late August or the first half of September and one on 29 September 1995 is the latest occurrence on record.

RED-THROATED PIPIT *Anthus cervinus*

One was seen and heard briefly in flight over the Sand Plant peninsula on 1 November 1995, on a day when a Richard's Pipit was also present among a large movement of Skylarks. This is the only record.

MEADOW PIPIT *Anthus pratensis*

Meadow Pipits are common, occasionally abundant, migrants at Marshside in both spring and autumn. Comparatively few are present in winter and the species nests in small and apparently declining numbers on both fresh and saltmarshes. Movements of small parties of Meadow Pipits in mid-February are usually among the first hints of spring passerine migration at Marshside. In most years passage continues intermittently through March and April and some very large flocks have been recorded, including 500 on 4 April 1982, 1,000 on 14 April 1991, and at least 2,500 grounded by a heavy downpour on 12 April 1993. Although data on breeding numbers in the past is lacking most regulars are agreed that there has been a perceptible decrease since the early 1980s, at least. It is likely that no more than 14 pairs bred over the entire recording area in 1999; six of these nests were on the saltmarsh.

Autumn passage of Meadow Pipits normally begins on a small scale in mid- to late August, gains momentum in September and continues intermittently until early November. The volume varies considerably from year to year but recent peaks include 500 on 17 September and 220 on 17 October 1993, 230 on 5 October 1994 and 2,800-plus moving south-southeast in 3.5 hours on 8 October 1995; since 1996 autumn movements at Marshside have been very light. Relatively few Meadow Pipits are present during the mid-winter period and ten in a day would be a very noteworthy total.

WATER PIPIT *Anthus spinoletta*

This elegant, elusive and sought-after pipit is a scarce and sporadic winter visitor and a passage migrant in spring. Regular wintering by one or more birds was a feature of the comparatively recent past; from 1978 to 1984 one or two birds were present during each winter, usually in the vicinity of Crossens Water Treatment Plant. Sightings were intermittent but birds were normally recorded between late November and mid-March and at least four were present during January-March in 1978 and 1984. In addition to overwintering birds there were several records of Water Pipits on spring passage in this period: one was seen on 23 March 1979, one on 15 April 1980, and an extraordinary total of eight on 1 April 1984. From 1985 to 1991 the species was totally absent from Marshside, but sporadic sightings recommenced in 1992 when a single bird was seen briefly by the Sand Plant lagoons on 6 and 30 December. Since then there have been at least 25 records, several in spring involving two individuals. All have occurred in the period November-May and most have unfortunately been of very brief duration.

ROCK PIPIT *Anthus petrosus*

The robust and familiar Rock Pipit is a winter visitor and passage migrant at Marshside. It is normally recorded in ones and twos, either flushed from the saltmarsh or else foraging quietly below the seaward side of the Marine Drive embankment after being dislodged from the vegetation by an unusually high tide. As in the case of its close relative, the Water Pipit, the recent history of the present species at Marshside appears to be one of sudden decline followed by partial recovery. Oakes (1953) describes the Rock Pipit as being no more than a casual wanderer to coastal areas of Lancashire in winter. An appreciable increase in numbers must have taken place in subsequent decades, as in the 1970s and early 1980s totals at Marshside occasionally reached substantial numbers. There were 49 on Crossens saltmarsh on 15 December 1974, 35 there on 8 February 1977, and 35 were counted over the whole of Marshside and Crossens in late December 1982. A sudden decrease appears to have commenced around 1984, and until 1990 wintering numbers were very low with a mere seven birds on 26 February 1989 the highest count during a six-year period.

Since then totals have recovered somewhat; three or four birds may be seen on a typical visit in mid-winter whilst at least 18 were flushed by a high tide on 23 December 1991 and up to a dozen were seen on 19 February 1995 and on several dates in January 1996. Numbers were again very low in 1997 and 1998, but 1999 saw another modest improvement.

Rock Pipits have been recorded at Marshside as early as 8 October and as late as 7 April in several recent years and there are exceptional records of one on 26 May 1995 and of two birds on 7 June 1980. It is probable that a proportion of spring and autumn records involve passage migrants from well outside the area; this is certainly true of an individual showing the characters of the Scandinavian subspecies littoralis present briefly on the Sand Plant peninsula on 10 March 1993, and another on the saltmarsh on 28 November 1999.

YELLOW WAGTAIL *Motacilla flava*

Of all our more familiar passerine migrants at Marshside none has suffered such a massive decrease in numbers over the past two decades as the delightful Yellow Wagtail. As recently as the early 1980s the species was a fairly common migrant in both spring and autumn. Numbers varied a good deal from year to year but day totals of 12 or 15 birds were quite normal in late April and early May, with a spectacular 300 on 3 May 1978 the highest count recorded. Totals in autumn were normally rather lower than in spring, perhaps reflecting reduced observer coverage at that season, but up to a dozen Yellow Wagtails were not infrequently seen on Stanley School field on August and early September mornings. As well as its being a regular migrant one or two pairs occasionally nested, usually at Crossens along the inland side of the Marine Drive. The last recorded breeding was in 1985, and since the mid-1980s numbers of Yellow Wagtails passing through Marshside have virtually collapsed.

Between 1986 and 1990 no count, either in spring or in autumn, exceeded single figures; in 1992 only one bird was recorded in spring and none at all in autumn. Totals in 1993 and 1994 remained at very low levels, eight and four birds respectively. Matters improved in 1995, however, and at least 15 Yellow Wagtails were recorded in spring and 24 in autumn. In 1996 the recovery continued, with about 40 recorded in spring and 11 in autumn, but numbers slumped again at both seasons in 1997 and 1998. 1999 produced yet another dismal spring, but the autumn total of 42 was the highest of the decade. All available evidence suggests that a widespread decline in numbers has taken place both nationally (Gibbons, Reid & Chapman, 1993) and within Lancashire (Jones, 1995).

There are two records of the nominate Blue-headed race at Marshside, males on 28 April 1985 and 28 September 1997, whilst an individual described in the Lancashire Bird Report as 'Black-headed' was seen on 5 June 1975 (Spencer 1976).

GREY WAGTAIL *Motacilla cinerea*

This most elegant of our wagtails is an uncommon autumn passage migrant at Marshside. One or two birds overwinter in most years, usually in the vicinity of Crossens Water Treatment Plant, but occurrences in spring are quite rare. The volume of autumn movements varies a great deal from one year to the next. In most years around 25 birds are recorded between mid-August and early November, but over 80 were seen in autumn 1995 with peaks of 18 on 13 September and at least 35 passing in 3.5 hours on 8 October.

PIED WAGTAIL *Motacilla alba*

The resident British and Irish race yarrellii is familiar at Marshside all year round as a breeding bird, passage migrant and winter visitor. Spring movements usually begin very early, in some years in early February, and continue intermittently until the beginning of April. Up to 40 birds may be seen on a typical day in mid-March. Migrant Pied Wagtails favour the short-grass areas of the fresh marsh and the edges of pools, and numbers tend to be highest in years when floods persist late into the spring leaving grassy islands surrounded by large exposed areas of mud. In recent years one or two pairs have nested regularly in the area of the Sand Plant compound and lagoons, and young were successfully reared every year from 1991 to 1998.

Autumn passage normally commences in mid- to late August and continues on and off until late October. Numbers are often higher than in spring and an exceptional 350 in 3.5 hours on 8 October 1995 and 75 on 20 September 1997 are the peak day totals in recent years. Up to ten widely-dispersed Pied Wagtails normally spend the winter at Marshside but occasional flocks are recorded, possibly hard-weather movements from other sites; 35 on 7 January 1992 is the largest such group in recent years.

The Continental race M. a. alba, the White Wagtail, is a fairly common passage migrant particularly in spring. Numbers recorded vary considerably from year to year, however, and there is strong evidence of an overall decrease over the past quarter century. Spring movements normally begin in early April, peak towards the end of the month and last until early May but ones and twos in the last week of March are not unusual, and in some years, as in 1995, a trickle of migrants continues up to about 20 May.

Some very impressive day-totals were recorded up to the early 1980s, including 150 on 14 April 1980, 200 on 20 April 1981 and 170 on 15 April 1982. Occasional large falls persisted up to about 1990 such as 150 on 21 April 1987 and 106 on 23 April 1989, but a gradual decline in numbers seems to have begun around 1983. By the early 1990s this decrease was obvious to all regular observers. No more than 60 were seen during spring 1991 and numbers crashed almost completely in 1992 with fewer than a dozen White Wagtails recorded in the entire season. Since then spring totals have recovered somewhat but they remain well short of those seen even a decade ago: 60 were recorded in 1993, 110 in 1994, 175 in 1995, and about 145 in 1996. After an ominous slump to an all-time low of three birds in spring 1997, 1998 saw a surprising rebound, with at least 230 recorded; there were 82 in 1999.

Autumn movements appear to have been consistently lighter than those in spring and consequently the fall in numbers is less easy to trace. There are, however, several day-counts of 30 or more dating from the early 1980s which may be compared with seasonal totals of 18 in 1990, 12 in 1991, zero in 1992, 17 in 1993, 15 in 1994, four in 1995, nine in 1996, 18 in 1997, nine in 1998 and one in 1999. Autumn White Wagtails are normally seen between mid-August and the end of September with the peak, such as it is, in early September.

WREN *Troglodytes troglodytes*

Until the early 1990s no-one appears to have paid any systematic attention to Wrens at Marshside. Consequently we know virtually nothing about the species's status even in the recent past, apart from a vague impression that there have usually been quite a few of them about. From the recent data it seems that Wrens are widespread winter residents in small numbers, found even on the inner edges of the saltmarsh, with up to 20 present in most years. They are somewhat more numerous as autumn passage migrants between late September and the beginning of November; recent peak counts include 45 on 20 October 1992, 55 on 10 October 1993 and 35 on 11 October 1994. Slight spring movements were noted for the first time in 1998 and 1999, peaking at 15 on 15 March 1998 and 16 on 1 April 1999. In recent years up to seven pairs have nested on Hesketh Golf Course with another one to three pairs on the Marine Drive embankment.

DUNNOCK *Prunella modularis*

The Dunnock is resident at Marshside with up to five pairs breeding on Hesketh Golf Course and another two or three along the Marine Drive embankment in most years. There is no evidence of any passage or irruptive movements.

ROBIN *Erithacus rubecula*

The profile of the Robin at Marshside is very similar to that of the Wren. Up to seven pairs nest on Hesketh Golf Course annually with one or two more along the Marine Drive embankment in some years. In addition the species is a late autumn passage migrant, occasionally in considerable numbers. Recent peak counts include 85 on 17 September and 78 on 10 October 1993, and 74 on 17 September 1998. Movements have normally ceased by mid-November but up to 15 birds are usually present in winter, most of them on the Golf Course but a few scattered along the Marine Drive and on the Sand Plant peninsula.

NIGHTINGALE *Luscinia megarhynchos*

A male that sang in shrubbery by the Crossens Water Treatment Plant compound from 20 to 24 May 1980 is the only record at Marshside of this rare southern overshoot.

BLUETHROAT *Luscinia svecica*

When one considers how seldom regionally rare passerines are recorded at Marshside it seems incredible that this normally-skulking vagrant has been seen on no fewer than four occasions since the early 1970s. A female was found near the Sand Plant lagoons during a very large fall of passerines on 5 May 1973, a male of the white-spotted race *cyanecula* was present briefly at the edge of Hesketh Golf Course on 13 March 1977, a male of the nominate Scandinavian race was in the north-west corner of M1 on 18 June 1978, and another *cyanecula* male provided excellent views to both local and visiting birders on the southern margins of M1 on 6 April 1995.

BLACK REDSTART *Phoenicurus ochruros*

The Black Redstart is a very scarce and irregular visitor to Marshside. Of nine records since the mid-1970s five have been in spring, all involving female birds, on 22-23 April 1978, 4 April 1988, 3 May 1994, 2 April 1996, and 28 April 1999. There have been two autumn records, single males on 27 October 1988 and 22 November 1997. In addition an immature male was present on the Sand Plant peninsula from 25 November 1986 to 11 January 1987, and a female from 18 December 1999 into 2000.

REDSTART *Phoenicurus phoenicurus*

The Redstart is an uncommon though regular spring migrant at Marshside. Numbers vary considerably from year to year but the species has been recorded in all but two of the 19 years since 1980. Totals in spring have tended to increase in the 1990s: there were two in 1990 and 1991, eight in 1993, seven in 1994, 11 in 1995, six in 1996, one only in 1997, a record total to date of 13 or 14 in 1998, and two in 1999. This may be in part a function of more intensive searching for spring passerine migrants, especially on Hesketh Golf Course, an area favoured by Redstarts. The passage period for this species is comparatively brief with the great majority recorded in the last week of April and the first half of May; one on the Golf Course on 29 March 1981 was a very exceptional occurrence. Redstarts are rare on autumn passage at Marshside with only six documented records of single birds in the last quarter century, all between mid-September and early October.

WHINCHAT *Saxicola rubetra*

Whinchats are fairly common spring passage migrants at Marshside, much scarcer in autumn. The first superbly-plumaged spring male is always keenly awaited; movements normally begin around 20 April and continue until the last week in May with a peak in early May. As with most other passerine migrants there is quite a large annual variation in numbers but at least 30 are seen in most years.

Substantial falls are occasionally recorded on easterly or south-easterly winds such as 44 on 9 May 1982, 31 on 11 May 1993, 20 on 2 May 1995 and 22 on 6 May 1997. Migrant Whinchats may turn up almost anywhere but the wire fences along the inland side of the Marine Drive embankment are particularly favoured perches. Prior to the 1990s sporadic breeding by one or two pairs took place, most recently in 1983 and 1989.

Autumn migrants are recorded in most years and passage is often protracted with earliest and latest dates during the last two decades of 7 July and 6 November, respectively, but most pass through in September. Totals in autumn are very much lower than in spring and no more than three or four are seen in an average year.

STONECHAT *Saxicola torquata*

Among passerines at Marshside the Stonechat holds a special place in the affections of the regular birders due at least in part to the fact that for several years in the early 1990s it seemed as though the species was doomed to disappear altogether from the site. Stonechats are uncommon migrants in early spring and autumn and up to four birds normally overwinter. Until 1987 breeding by one or two pairs took place about every third year.

Spring movements are observed between mid-February and early April and numbers vary considerably. Between five and ten have been seen in most recent years but at least 35 passed through in spring 1994 with peaks of nine on 9 and 17 March.

Autumn passage is also very variable, but normally up to eight birds are seen between late September and the middle of November. 1994 with up to 20 birds, 1995 with 18 and 1999 with at least 22 were exceptionally good autumns. Overwintering Stonechats favour the vicinity of the Sand Plant peninsula and the road edges and the birds can be remarkably elusive, but three or four have been present between late November and at least the end of January in most recent winters.

The recent and still precarious recovery in the fortunes of this most appealing passerine at Marshside is all the more gratifying in the light of a virtual collapse in numbers during 1989-91 when only one individual was fleetingly recorded in the whole three-year period. Totals of migrants if not yet of winter birds are about as high in the late 1990s as they were in the late 1970s and a resumption of nesting is hoped for, if not quite expected. The species has declined dramatically as a breeding bird in Lancashire during the 1990s (Jones, 1995).

WHEATEAR *Oenanthe oenanthe*

This is the spring migrant par excellence and few other birding experiences in the entire year can compare with that first glimpse of a perfect male Wheatear on a blustery March morning. The Wheatear is common on spring passage at Marshside; an average of about 210 per year has been recorded during the 1990s. Movements extend from the middle of March to about 25 May in most years, with peak counts anytime between mid-April and about 10 May. A variable proportion of birds in late April and May show the characters of the larger and brighter Greenland race leucorhoa.

Numbers in spring may have decreased somewhat over the last two decades. In the mid- to late 1990s peak day totals of 15 to 20 are typical, and 43 on 2 May 1995 and 46 on 20 April 1996 were fairly exceptional. This may be compared with fairly frequent day-counts of 40 or more in the late 1970s and 1980s and a spectacular 100-plus on 23 April 1980. A pair attempted to breed in a Rabbit burrow on the Sand Plant peninsula in 1993 but abandoned the site after two weeks; this is the only instance of nesting recorded in the last two decades.

Movements of Wheatears in autumn are generally much lighter than those in spring. Very occasionally the first migrant is seen before the end of June but passage normally takes place between late July and the end of September with odd stragglers up to the end of October. Fewer than 30 autumn migrants are seen in a typical year with a peak day count of seven or eight usually recorded sometime in late August or early September. 40 on 24 September 1980 is by far the highest day total in recent decades.

RING OUZEL *Turdus torquatus*

This attractive upland thrush is a scarce but possibly increasing spring migrant at Marshside, very rare in autumn. There are fewer than 35 records since the mid-1970s and quite a few years have passed without a single sighting; in the early 1990s there were only two spring occurrences, on 23 April 1990 and 31 March 1993. The species has been almost annual in more recent springs, however, with two each in 1996 and 1997 and an unprecedented seven or eight in 1998. Records span the period from late March to mid-May but the great majority have been in April. Most Ring Ouzels are seen along the Marine Drive embankment or by the seaward edge of Hesketh Golf Course, and almost all pass through very quickly. The only records in autumn during the last quarter century are of single birds on 11 September 1992 and 24 September 1997.

BLACKBIRD *Turdus merula*

Blackbirds are resident at Marshside with around six pairs nesting on Hesketh Golf Course and up to three pairs on the Marine Drive embankment in a typical year. In addition small numbers pass through in late autumn with other migrant thrushes; peak counts in recent years include 20 on 24 October 1993, 22 on 13 October 1996 and 35-plus on 19 November 1998.

FIELDFARE *Turdus pilaris*

Fieldfares are recorded annually on autumn passage at Marshside, mainly in October, but numbers vary greatly and the vast majority of birds are seen moving high overhead with only a few pausing to feed, usually on Hesketh Golf Course. 700 flying over in 3 hours on 16 October 1994 is the largest movement recorded in recent years. In winter Fieldfares are occasional visitors, usually in small flocks, to the Golf Course and the fresh marsh. These occurrences have become markedly fewer since the early 1980s. A few spring migrants are recorded in most years, mostly in late March and April, with three on 6 May 1978, one on 27 April 1991 and another on the same date in 1997 the latest in recent decades.

SONG THRUSH *Turdus philomelos*

Like the Blackbird the Song Thrush is a breeding resident and autumn migrant at Marshside and up to ten birds overwinter in most years. Nesting numbers are much lower than for the Blackbird, however. Song Thrushes do not breed on the marsh proper and there are no more than two pairs on Hesketh Golf Course in an average year. Unfortunately, comparative data from the 1970s and 1980s is lacking so one cannot assess whether a decline in breeding numbers has taken place. Numbers on autumn passage are seldom large and 50 on 26 September 1981, 35 on 24 October 1993 and 25 on 13 November 1994 are the highest day totals recorded.

REDWING *Turdus iliacus*

Like its relative and fellow-traveller, the Fieldfare, the Redwing is a regular migrant at Marshside in autumn and an occasional visitor in winter and spring. Numbers for the present species also fluctuate widely from one autumn to the next and 570 moving overhead with Fieldfares on 16 October 1994 is the largest documented count in recent decades. A few casual visitors are seen, usually on the Golf Course, in most winters although numbers are occasionally greater in hard weather. Totals in winter were, however, consistently rather higher in the late 1970s and early 1980s. There are several recent records of late spring migrants with 25 on 26 April 1983 and single birds on 4 and 19 May 1993 and on 8 May 1994 the most notable.

MISTLE THRUSH *Turdus viscivorus*

One or occasionally two pairs of Mistle Thrushes breed annually on Hesketh Golf Course and parents and offspring may frequently be seen feeding out on the fresh marsh in late spring and summer; in 1999 three pairs nested for the first time, all successfully. There is also evidence of occasional migratory or irruptive movements in autumn and winter with a few Mistle Thrushes apparently accompanying other thrush species. A flock of 13 over M1 on 24 October 1993 and nine together over the Golf Course on 20 October 1998 are the most recent of these occurrences.

GRASSHOPPER WARBLER *Locustella naevia*

This notorious but atmospheric little skulker is a scarce passage migrant, mainly in spring. Almost all the spring records are of reeling males and the number of birds that move silently through each year can only be guessed at. One to three are recorded in most years with most turning up on Hesketh Golf Course, in the SSSI ditch, or along the Marine Drive embankment, but only very rarely will a bird remain for more than a single day. Up to seven widely-dispersed migrants on 30 April-1 May made 1997 the best spring yet for this species. Most Grasshopper Warblers occur between the last week of April and the middle of May but odd migrants have been recorded as early as 14 April and as late as 20 May in recent decades. A single pair bred on the Golf Course in 1981.

Singles on the Marine Drive embankment on 17 September 1993 and on the Golf Course on 13 August 1996 are the only recorded occurrences in autumn, although Grasshopper Warblers are at their most unobtrusive at this time of year and quite a few others have no doubt passed through unobserved.

SAVI'S WARBLER *Locustella luscinoides*

A reeling male was present below the Marine Drive embankment on M1 for several hours on the afternoon of 28 April 1977. This was the first of six records of this southern European reedbed specialist in Lancashire up to the end of 1999.

SEDGE WARBLER *Acrocephalus schoenobaenus*

The exuberant if somewhat discordant song of the Sedge Warbler is one of the most characteristic sounds of Marshside in late spring and early summer. There is a variable breeding population with pairs distributed along the inland side of the Marine Drive embankment from Crossens Water Treatment Plant to Hesketh Road, as well as on the edge and even in the interior of the Golf Course. Breeding numbers increased fairly steadily over the two decades up to the mid-1990s and an annual average of 27 pairs nested in the five seasons 1991-95. After 1996 there was a sudden and worrying decline, however; only 12 pairs nested in 1997 and seven in 1998. A partial recovery to 16 breeding pairs in 1999 was reassuring, but we may not be out of the woods yet. Success rates of the pairs that do nest seem to be reasonably high with plenty of juveniles about in July and early August.

The first Sedge Warbler is usually seen, or more often heard, around 20-23 April; one on 29 March 1997 was the earliest ever recorded. Even the first migrants tend to sing lustily and it is difficult to determine on any given visit which individual birds are merely on passage and which have already established a breeding territory.

Autumn movements are not conspicuous at Marshside and it is impossible to determine whether birds glimpsed skulking in the rank vegetation of August and early September are migrants, local nesters or their offspring. Whichever is the case, very few Sedge Warblers are recorded after the middle of September.

REED WARBLER *Acrocephalus scirpaceus*

Since the late 1980s a number of small breeding pockets of Reed Warblers have become established in South-West Lancashire, and Marshside has benefited from this range expansion. Prior to 1990 the Reed Warbler was a very scarce passage migrant at Marshside, sometimes on unusual dates as in the case of the two very late (or very early) birds on 10 July 1983, or stranger still the individual trapped and ringed on 16 December 1984. In 1990 a pair bred successfully in a small patch of reeds below the Marine Drive embankment on M1. Although a singing male was present at the same spot for several days in mid-May 1991 nesting did not take place and there were no records at all in 1992. In the following year, however, two pairs bred successfully by the M1 embankment; this was repeated in 1994 and a single pair reared at least three offspring in 1995. Although a few spring and autumn migrants were recorded in 1996 and 1997 there were no attempts at nesting; in 1998, however, three pairs bred, two of them for the first time in the SSSI ditch. Four pairs, all of them successful, in 1999 ended the decade on a very positive note.

BARRED WARBLER *Sylvia nisoria*

An immature bird was present in the shrubbery by Crossens Water Treatment Plant on 16 September 1967. This scarce but regular autumn migrant on east and south coasts is a rare visitor to North-West England, so it is not surprising that this is our only record.

LESSER WHITETHROAT *Sylvia curruca*

According to Oakes (1953) the Lesser Whitethroat had been a sparse breeder in South Lancashire prior to the Second World War but had since declined. A slight increase in breeding numbers seems to have taken place during the 1980s and 1990s, but the species remains very thinly distributed in the County. At Marshside Lesser Whitethroats are scarce passage migrants, mainly in spring. Most are recorded on Hesketh Golf Course and although a few singing males have stayed for several days breeding has never been suspected; records extend from 27 April to 1 June.

Only two spring migrants were recorded during the whole of the 1980s but the rate of occurrence greatly increased in the 1990s due in part, perhaps, to more systematic coverage of the Golf Course. There was one in 1991, another in 1994 and an unprecedented three in 1995, including two birds on 2 May. 1995 also saw the first autumn migrants recorded for the last two decades at least, with two at the edge of the Golf Course on 26 July and another by the M1 embankment on 20 August. Two spring migrants in 1996, six in 1997, one in 1998 and two in spring plus another in autumn 1999 have maintained the pattern of increasing occurrences, and in 1996 and 1997 a pair nested in brambles by the Marine Drive near Crossens Roundabout, just outside our recording area.

WHITETHROAT *Sylvia communis*

Alongside the song of the Sedge Warbler the cheerful stutter of newly-arrived Whitethroats is a characteristic sound of Marshside in spring. The Whitethroat is a widely-distributed and successful breeder, nesting all along the Marine Drive embankment; the species's stronghold, however, is on Hesketh Golf Course where up to 13 pairs have bred in recent seasons. The number of Whitethroats nesting at Marshside-Crossens as a whole has increased fairly steadily from maxima of around 10-12 pairs in the late 1970s and early 1980s to an annual average of 19 pairs during the 1990s.

Spring movements normally get under way in the last week of April and continue until the end of May with peaks in early May. Large falls are few but 25 were seen on 11 May 1993, 20 on 1 May 1994 and 28 on 1 May 1997.

As with other migrant passerines it is difficult to distinguish genuine autumn passage birds from local Whitethroats, but most seem to move through or depart by late August. One on 1 October 1980 is the latest record in the past quarter century.

GARDEN WARBLER *Sylvia borin*

This unassuming but pleasant songster is an uncommon annual passage migrant at Marshside; breeding has occurred on Hesketh Golf Course during the 1990s. Spring migrants are recorded between the last week in April and the middle of June with the great majority in the first half of May. Five or six are seen in a typical spring but five were present together on 8 May 1982 and four on 12 May 1993; there was a record total of 18 spring migrants in 1998. Most Garden Warblers are recorded on Hesketh Golf Course but a few have also been found along the Marine Drive embankment. A single pair bred on the Golf Course each year between 1991 and 1993. In 1995, after a late influx involving at least three singing males in mid-June two pairs successfully reared young, both on the Golf Course.

Migrant Garden Warblers are scarcer in autumn than in spring but two or three have been seen in most seasons since 1990; there were seven in autumn 1996 and four in 1999. Most occur in the second half of August or the first week of September, and one on 13 October 1998 is by far the latest on record.

BLACKCAP *Sylvia atricapilla*

The Blackcap is an uncommon passage migrant in both spring and autumn, and a very recent breeding colonist. Passage birds are recorded both on the Marine Drive embankment and on Hesketh Golf Course and there has been a noticeable increase in numbers during the 1990s. Spring movements extend from mid-April to the beginning of June with most occurrences in the first half of May; totals have increased steadily during the 1990s, to a record 23 in 1999.

In 1994 a single pair bred successfully on Hesketh Golf Course, this foothold increasing to two pairs in 1995-96, three in 1997-98 and five in 1999. This has coincided with the apparent abandonment of the area by our small breeding population of Garden Warblers; there is some evidence that Blackcaps, which arrive earlier in spring, tend to drive Garden Warblers from their territories (Gibbons et al, 1993).

Autumn movements of Blackcaps extend from the middle of August to the end of November. These have also become generally more conspicuous in the 1990s with seven recorded in 1992, eight in 1993, six in 1994, eight in 1995 and 1996, nine in 1997 and at least 20 in 1999. As in spring almost all are seen on the Marine Drive embankment or on the Golf Course; most are recorded in September and six on 28 August and 26 September 1999 is the largest day total. There are no mid-winter records.

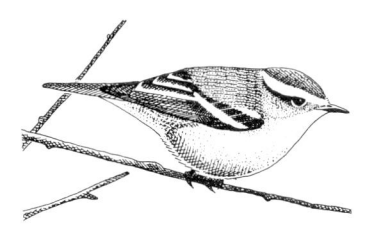

YELLOW-BROWED WARBLER *Phylloscopus inornatus*

One on Hesketh Golf Course on 29 September 1997 was our long-awaited first record of this scarce Siberian wanderer.

WOOD WARBLER *Phylloscopus sibilatrix*

This frequenter of upland oakwoods is a very scarce though increasingly regular spring transient at Marshside. Most are recorded on Hesketh Golf Course although one or two have been seen fleetingly on the Marine Drive embankment. Since the early 1980s there have been records in 1985, 1986, 1989, 1993, 1995, 1997, 1998 and 1999. Almost all have been seen between late April and mid-May and two on 4 May 1985 and on 12 May 1993 are the only multiple occurrences. The great majority of records are of singing males and only two have remained for more than a single day. There are no autumn records.

CHIFFCHAFF *Phylloscopus collybita*

The Chiffchaff is a fairly common spring passage migrant at Marshside, somewhat scarcer in autumn. Occasional breeding by a single pair has taken place in recent years and there are a quite few mid-winter records. Spring movements are protracted, with the first migrants appearing on the Golf Course or along the Marine Drive embankment as early as the beginning of March in some years and occasional birds still passing through in late May. Numbers vary a little from year to year with fewer than ten in the cold, bleak spring of 1991 but over 25 in 1993, 35 in 1994 and an average of 34 birds annually during 1995-99. Eight on 22 April 1993, ten on 22 April 1994 and six on 2 May 1995 are the highest day totals in recent years. In 1991 and 1993 a pair nested, apparently successfully, on Hesketh Golf Course.

Occasional birds are seen in July and early August, presumably as a result of post-breeding dispersal from nearby sites, but autumn passage proper does not normally commence until late August and continues until the end of October. An average of 20 birds has been recorded each autumn during the 1990s; 1997 was exceptional, with a total of 44 and a peak of 14 on 28 September. Chiffchaffs showing the characters of the Scandinavian race abietinus have been seen on 14-15 October 1985 (two), 15 November 1987, 27 November 1997, and from 17 December 1998 into 1999.

There have been at least 15 records of Chiffchaffs in mid-winter during the last quarter century. Hesketh Golf Course and the willows by Crossens Sewage Works are favoured sites and at least three were present on the Golf Course during hard weather in late December 1995.

WILLOW WARBLER *Phylloscopus trochilus*

The Willow Warbler is a common passage migrant in both spring and autumn and a regular breeder on Hesketh Golf Course. Migrants are recorded in almost all habitat types but are most numerous on the Golf Course and along the Marine Drive embankment. Spring movements normally commence in the second week of April and continue intermittently until mid-May, but first arrival dates are greatly dependent upon prevailing weather patterns and may be as early as 28 March or as late as 22 April. Numbers of migrants also appear to be determined by the weather; in each of the inclement springs of 1991 and 1992 fewer than 40 were recorded in contrast to totals of 150-200 in each of the seasons from 1993 to 1996, over 250 in 1997, a record total to date of 345 in 1998, and 117 in 1999.

Suitable conditions occasionally produce spectacular falls with birds flitting about in every available shrub and bramble-patch: 80 were present on 3 May 1986, 66 on 12 May 1993 and an unprecedented 94 on 28 April 1998. In each year since regular censusing began in 1989 between four and eight pairs of Willow Warblers have bred on the Golf Course.

Autumn passage tends to begin and finish early with movements usually commencing at the end of July and petering out by mid-September. Totals are normally lower than in spring although reduced observer coverage in autumn may in part account for this. Peak numbers are recorded in the first half of August and 32 on 11 August 1996 is the highest count in recent years. One on 4 October 1984 is by far the latest record in the last quarter century.

GOLDCREST *Regulus regulus*

Small, fast-moving groups of Goldcrests are a characteristic feature of autumn passerine migration at Marshside. They are liable to turn up almost anywhere, even on the sparsely-vegetated tip of the Sand Plant peninsula. Movements usually begin at the end of August and continue on-and-off until early November, with peak day totals in recent years of 55 on 26 September 1993, 60 on 11 October 1994 and 63 on 5 October 1997. The species is quite rare in mid-winter although one or two are occasionally to be found with parties of tits on Hesketh Golf Course in hard weather.

Spring passage normally occurs from mid-March to late April, and numbers though lower than in autumn, seem to have grown during the 1990s. 18 on 14 April 1994, 17 on 31 March 1998 and 32 on 18 March 1999 are the highest day counts in recent decades. A few late migrants are sometimes seen well into May but there is no evidence of breeding.

FIRECREST *Regulus ignicapillus*

A male was present on the Marine Drive embankment of M1 for most of 3 May 1997, and another, also probably a male, was on the Golf Course on 21 November 1999. These are the only documented records.

SPOTTED FLYCATCHER *Muscicapa striata*

This late-arriving migrant is an uncommon but annual spring transient at Marshside. Numbers vary considerably from year to year and most are seen on Hesketh Golf Course. All the recent sightings have been at the very end of April or in May, and between four and seven are recorded in a typical season. Six on 8 May 1988 and on 25 May 1997, at least 20 on 11-12 May 1993 and five on 13 May 1998 are the only substantial multiple occurrences in recent decades. In 1993, following an exceptionally heavy passage a pair bred successfully on the Golf Course, but there have been no subsequent attempts.

Spotted Flycatchers are scarcer in autumn, but records seem to have increased during the late 1990s; there was one in September 1993, three between late August and early October in 1996, five in August-September 1997 and five in September 1999. One on 10 October 1978 is the latest on record.

PIED FLYCATCHER *Ficedula hypoleuca*

The Pied Flycatcher is a very scarce migrant at Marshside with only nine recorded in spring and three in autumn during the last quarter century. Birds have been seen on Hesketh Golf Course as well as on the Marine Drive embankment and most visits have been very fleeting. Spring migrants have been recorded between 28 April and 20 May whilst all three autumn birds were found in mid to late September. 1995, with three birds in spring and another in autumn, was wholly exceptional.

BEARDED TIT *Panurus biarmicus*

One was seen on 18 December 1977 during a small influx of this irruptive species into South-West Lancashire, which also brought individuals to Martin Mere WWT on 3 December and to Mere Brow on 14 December.

LONG-TAILED TIT *Aegithalos caudatus*

Small parties of Long-tailed Tits may be encountered on Hesketh Golf Course at any season and occurrences seem to have become more frequent in recent years. In 1994 a pair bred successfully for the first time and this was repeated in 1995-97. Although display by a pair was observed in spring 1998 nesting did not take place; however, two pairs bred successfully in 1999.

There is very little clear evidence of any migratory movements and the species is hardly ever recorded beyond the Golf Course boundary; up to 15 along the verges of Marshside Road on 20 October 1998 was the first such occurrence for many years.

COAL TIT *Parus ater*

The Coal Tit is a very scarce and irregular late autumn migrant at Marshside with fewer than a dozen records during the past quarter century. Most have been seen on Hesketh Golf Course, and a flock of 28 on 24 September 1997 is by far the highest number recorded. Two by the Sand Plant lagoons on 1 May and one on the M1 embankment on 28 May 1997, and yet another by the lagoons on 15 May 1999 are most exceptional records.

BLUE TIT *Parus caeruleus*

Blue Tits are fairly common at Marshside all year round, particularly on Hesketh Golf Course, although they are also frequently seen on the Marine Drive embankment especially during autumn dispersals and passage movements. These normally commence on a small scale in mid-August and up to 40 birds may be recorded on a good day between mid-September and the beginning of November. 80 on 19 October 1993 is the highest count in recent years. Blue Tits breed in fair numbers on the Golf Course; it is likely that at least seven pairs nest in a typical spring.

GREAT TIT Parus major

Great Tits in small numbers often accompany parties of migrant Blue Tits at Marshside in autumn. Up to 20 may be seen on Hesketh Golf Course on a good day in late September or October although very few are recorded on the Marine Drive embankment. In addition the species is regularly present on the Golf Course in winter, and nests annually; the recent erection of nest-boxes on the Golf Course has produced an increase from five breeding pairs in 1994-98 to about nine in 1999.

RED-BACKED SHRIKE Lanius collurio

An adult male was found in the south-west corner of M1 on the afternoon of 16 May 1996 and relocated briefly on Hesketh Golf Course later in the evening.

GREAT GREY SHRIKE Lanius excubitor

One was present for only a few minutes on Hesketh Golf Course around midday on 2 April 1982. This is the only record.

JAY Garrulus glandarius

Dispersive or irruptive movements in autumn and winter occasionally bring one or a few Jays to Marshside, and visits seem to have become more frequent in the late 1990s. As one might expect, most are seen on Hesketh Golf Course but in the 1994 and 1996 autumns several birds spent some hours in bushes along the Marine Drive embankment. Since March 1998 there have been several visits to the Golf Course in spring; in 1999 two birds were seen on a number of occasions in May and June.

MAGPIE Pica pica

At time of writing Magpies are abundant and obtrusive everywhere at Marshside, even on the inner edge of the saltmarsh. Flocks of up to a dozen birds are common on Hesketh Golf Course at any time of year while others haunt the shrubbery along the Marine Drive embankment. Nesting pairs totalled three along the embankment and another four to five on the Golf Course in 1997 and 1998; the total increased to ten pairs in 1999. Although the Magpies are generally viewed with disfavour by birders their actual impact as nest predators on other bird species is difficult to estimate and may in fact be slight.

The Magpie's recent history at Marshside is also quite unclear as there are few if any references to the species in field notebooks from the 1970s and early 1980s. Ones and twos were regularly seen on the Golf Course around 1977 but visits to the seaward side of the area do not seem to have become regular until about 1982, and first breeding on the Marine Drive embankment probably dates only from about 1988. At the end of the 1990s nesting numbers have probably reached the maximum carrying capacity for the site.

JACKDAW *Corvus monedula*

Jackdaws are increasingly regular visitors to the fresh marshes, especially in autumn and winter. The species's nesting population in Southport as a whole has grown substantially since the mid 1990s; there are quite a few pairs in the Marshside/Churchtown area and it is likely that our visitors belong to this population. In late December 1999 the feeding flock on M2 reached a peak of 47 birds. A flock of 52 high over Hesketh Golf Course on 19 October 1997 is the only recent evidence of any migratory or irruptive movements.

ROOK *Corvus frugilegus*

Rooks are rather scarce in coastal South-West Lancashire as a whole so it is not surprising that the species is only a rare and sporadic visitor to Marshside. Most of the ten or so records since 1980 have occurred either in March/April or in September/October and only two involved more than a single bird, when four passed over together on 24 April 1988 and three on 27 September 1998.

CARRION CROW *Corvus corone*

The resident nominate race is regularly seen in small numbers at Marshside on both fresh and saltmarshes. Three or four birds together is around the norm and 18 on the saltmarsh on 8 October 1995 was quite exceptional. A single pair bred on Hesketh Golf Course on several occasions in the early 1990s.

The Hooded Crow *C. corone cornix* is a very scarce and irregular visitor in autumn, winter and early spring, mainly to the shore and saltmarshes. One seen on several dates in late March and April 1990, and one on 18 October 1998 are the most recent occurrences.

RAVEN Corvus corax

One was shot 'at Marshside' in 1914 (Oakes, 1953). Until very recently this was the only record, but two birds visited both fresh and saltmarshes on several dates between 20 February and 26 April 1999.

STARLING Sturnus vulgaris

Starlings are very common at Marshside in all habitat types but are seldom paid much attention by birders except when an unlucky individual falls prey to a Merlin or a Sparrowhawk. The species breeds abundantly in nearby residential areas and large numbers regularly feed on the Golf Course as well as on both fresh and saltmarshes at all times of the year. The sudden invasion of raucous dull-brown juveniles and their harrassed parents in late May is one of the definitive signs that spring is all but over.

Starling flocks are at their largest in the winter months when they presumably include many immigrants from northern Europe. There is little count data but it is likely that several thousands are present at peak times; 10,000 over the saltmarsh on 11 November 1998 is by far the highest total recorded.

HOUSE SPARROW Passer domesticus

Small parties of House Sparrows are occasionally present around the Sand Plant and its adjacent lagoons, as well as along the inland edges of the fresh marshes nearest to residential areas, and on Hesketh Golf Course. Most sightings are in late spring and summer. Intermittent nesting by two or three pairs took place in the Sand Plant buildings until the mid-1990s, and tree-nesting by several pairs took place on Hesketh Golf Course in 1990.

TREE SPARROW Passer montanus

As recently as the mid-1980s Tree Sparrows were regular visitors in small numbers, mainly in winter. Most were seen on Crossens Inner and saltmarsh and along the Marine Drive embankment immediately south of the Water Treatment Plant; 45 on 17 February 1977 and 50 on 13 February 1985 were the largest numbers recorded. After about 1986 the species suddenly became very scarce and erratic in its appearances, and during the 1990s several successive years have passed without a single record. In the early spring of

1995 there were several sightings of single Tree Sparrows near the Sand Plant and on Hesketh Golf Course. This was followed by a number of records in the latter half of the year, including parties of eight on 20 August and six on 31 October; this pattern of mainly late-year occurrences was repeated in 1996, 1997 and 1999.

CHAFFINCH *Fringilla coelebs*

Chaffinches are normally seen in small numbers at Marshside during autumn passage movements, especially in October. Most are recorded flying overhead or briefly feeding on Hesketh Golf Course, and 15 on 24 October 1993 and 25 on 11 October 1994 are the largest numbers seen in recent years.

At other seasons the species is fairly scarce and is virtually confined to the Golf Course where a small breeding population has become established since the mid-1990s; this reached five pairs in 1998. Small parties, usually fewer than ten birds, may occasionally be seen during spells of hard winter weather.

BRAMBLING *Fringilla montifringilla*

The Brambling is an irregular autumn migrant and winter visitor to Marshside. In the fairly recent past some very large winter flocks were recorded but the species has become very scarce and sporadic since the mid-1980s. Most were seen on the saltmarsh between early December and mid-February and peak counts include over 300 in mid-January 1979, 130 on 28 December 1981 and up to 500 from mid-January to late February 1984. Since the mid-1980s there have been very few sightings in winter although a few are still occasionally recorded on passage in late autumn; two over M1 on 24 October 1996, two over Hesketh Golf Course on 17 October 1997, and a total of four fly-over individuals in September/ October 1999 are the most recent occurrences.

GREENFINCH *Carduelis chloris*

Greenfinches are common and conspicuous at Marshside throughout the year. A few pairs normally nest on Hesketh Golf Course and on the Marine Drive embankment, and small feeding-parties may be seen along the embankment or on the peninsula on almost any visit. Flocks in autumn and winter are often quite sizeable: 300 were present in January 1985 and 200 in January-February 1986. Totals in more recent years have been rather smaller, with 130 in late November 1994 and 125 in late December 1995 the highest counts recorded.

GOLDFINCH *Carduelis carduelis*

The Goldfinch has increased substantially in numbers at Marshside during the last two decades, particularly in winter when formerly it was quite scarce. Three or four pairs now breed on Hesketh Golf Course in a typical spring and since 1995 at least one pair has nested annually in bushes on the M1 embankment. In late summer and autumn large flocks congregate on the fresh marshes, especially M1, to feed on thistle-heads. Over 100 were present in early September in 1988, 1989, 1995 and 1997 and a record total to date of 450 was recorded on several dates in mid-September 1998. Numbers between November and March are normally much lower, but up to 50 were seen regularly in mixed flocks with Linnets and Greenfinches in the vicinity of the Sand Plant during the 1995-96 and 1996-97 winters.

SISKIN *Carduelis spinus*

Siskins are uncommon but fairly regular autumn and spring passage migrants at Marshside; occurrences seem to have increased in frequency during the 1990s. Most are recorded in small flocks on, or more usually, over Hesketh Golf Course and the majority pass through very quickly. Numbers tend to be highest in autumn; records extend from mid-September to the beginning of November, and 35 on 19 October 1993 and 32 on 8 October 1995 are the largest day totals in recent years. Spring movements are less predictable; most occur in March and early April and 48 on 12 March 1997 is the highest day count recorded. In addition to migrant birds there are also a few scattered records of small parties of Siskins on the Golf Course during spells of hard winter weather.

LINNET *Carduelis cannabina*

Although Linnets are present at Marshside throughout the year and the population is swelled by an annual influx of winter visitors, there is no doubt that this familiar and appealing little finch has suffered a substantial decline in numbers since the early 1990s. During the late 1970s and the 1980s breeding totals fluctuated fairly widely from season to season but on average around 30 pairs nested, mainly along the ungrazed margins of the fresh marshes and on the inner portion of the saltmarsh. Winter flocks on the saltmarsh and shore were often spectacular. 1,000 were estimated on 19 January 1985 and over 400 were present in December of the same year; there were at least 700 in December 1988 and over 300 as recently as 1 February 1991. By 1992, however, the total number of nesting pairs was down to 12 or 13 and the winter flock had crashed to a mere 65 in January and 45 in December.

A modest recovery in wintering numbers was observed from late 1994 when about 90 were present; an influx in November 1995 brought the total at the year's end to 130-plus, seen mainly around the Sand Plant area. A major hard-weather movement in the last days of 1996 brought numbers on the saltmarsh nearly back to the levels of a decade earlier: 400-plus were present on 29 December, and up to 350 remained into January 1997. Wintering numbers were very low again in 1998 and 1999, however. Meanwhile the breeding population continues to dwindle; no more than five pairs nested in spring 1999.

TWITE *Carduelis flavirostris*

The Twite shares with Bean Goose and Goshawk the unenviable status of being one of Marshside's most illusory species; year in, year out Twites are confidently but erroneously identified among the Linnet flocks and added to year lists by visitors who will accept no expressions of doubt from more cautious onlookers. In the case of the Twite in particular the problem arises not only from the variability of female and immature plumages in Linnets but also from the fact that Marshside has a long-standing reputation as a regular wintering-site for Twites. In reality this image is somewhat outdated. Although moderate influxes took place during the 1995-96 and 1996-97 winters, for most of the 1990s numbers have been only a shadow of those regularly recorded prior to about 1989 and in some years the species has been entirely absent. During the 1970s and most of the 1980s Twites appeared annually on the saltmarsh and along the Marine Drive embankment around the middle of November and the flock, or sometimes several smaller groupings, would remain until early March. Up to 40 birds were present in a typical season although numbers were sometimes much higher: 70-plus in December 1981, 120 in late January 1984 and 75 as recently as January-February 1987. Smaller numbers, presumably migrants, were occasionally recorded in October and in April and May.

After the 1989-90 winter Twites became much more erratic in their appearances and the numbers involved greatly declined, to such an extent that a brief stay by up to eight birds on the peninsula in November-December 1994 was greeted with general rejoicing. Matters improved further with an influx of up to 30 in mid-December 1995, some of which remained into 1996. A flock of up to 60 birds in breeding plumage was present on both salt and fresh marshes during most of April 1996, and the biggest arrival for over a decade occurred during hard weather in December 1996 when up to 100 Twites accompanied Linnet flocks on the saltmarsh and along the shore. Hopes were high of a sustained return to the former days of plenty, but the only records since are of two birds on 31 October 1997 and three on 19 November 1999, all presumably on passage.

LESSER REDPOLL *Carduelis cabaret*

Redpolls nested annually in fluctuating numbers on Hesketh Golf Course from at least 1985 to 1998. A lack of systematic coverage of that area prior to the mid-1980s precludes any estimate of when breeding commenced, although there are scattered records of birds present in the nesting season as far back as 1977. Redpolls are seldom seen on the Golf Course between August and March although a flock of 40 was recorded in January 1990.

From 1985 to 1989 there were no more than three nesting pairs in any season; this increased dramatically to seven or eight pairs in 1990-91 and to at least 12 pairs in 1992. Since then there has been a return to more modest totals: six pairs in 1993, seven in 1994, no more than four in 1995, two or three in 1996 and 1997 and only one or two in 1998; nesting did not occur in 1999. Redpolls are rarely encountered beyond the Golf Course boundaries although there are a few records of overflying migrants in late autumn and early spring.

BULLFINCH *Pyrrhula pyrrhula*

This retiring inhabitant of scrubby woodland is exceedingly rare at Marshside in spite of the presence of apparently suitable habitat on Hesketh Golf Course. Three on the Marine Drive embankment in a general fall of migrants during heavy rain on 12 April 1993, a single bird on the Sand Plant peninsula on 25 November 1995 and another on the Golf Course on 26 October 1997 are the only records I could locate since the mid-1970s.

LAPLAND BUNTING *Calcarius lapponicus*

Lapland Buntings are much sought-after by visitors to Marshside in winter but very seldom seen. Although it is possible that small numbers are regularly present in most years on the vast and unfrequented expanses of the saltmarsh, it is more likely that the species is no more than a very scarce and irregular visitor on autumn passage and in winter. There have been records of Lapland Buntings in 15 of the 30 years since 1970 but in the great majority of cases sightings have amounted to no more than a brief glimpse of one or two birds. Most records are of birds on, or more usually flying over, the saltmarsh or the Marine Drive embankment between early October and late February.

The rate of occurrences has clearly declined since the mid-1980s; 1979-80 is still vividly remembered as the vintage winter for Lapland Buntings at Marshside. Two on Crossens saltmarsh in late December quickly increased to over 30 by late January; several birds remained until at least 1 March, providing some excellent views of the varied plumages of this exciting and enigmatic passerine. A single bird at Crossens on 24 February 1993, another on 26 December 1995 and one calling over M2 on 2 October 1998 are the most recent occurrences.

SNOW BUNTING *Plectrophenax nivalis*

Oakes (1953) describes the Snow Bunting as a regular winter visitor to the coast of Lancashire, although he considered it to be less plentiful than it had been earlier in the century, citing a record of 200 at Hundred End on 8 February 1929. As recently as 1976 Snow Buntings wintered annually along the seashore between Southport Pier and Formby Point and flocks of 20 to 30 birds were not unusual. Visits to Marshside, though less regular, were still quite frequent. Up to six spent much of the 1976-77 winter in the vicinity of the Wildfowlers' car park and 12 birds were regularly seen between January and March 1979. As recently as January 1984 over 40 Snow Buntings spent several days on Crossens saltmarsh but by the late 1980s the species had become very scarce. None were recorded in 1985 or 1989, and two birds in January 1987 was the only multiple occurrence. During most of the 1990s the situation deteriorated further; a single bird present briefly on the Sand Plant peninsula on 2 March 1994, one on Crossens Inner on 5 April 1996 and another along the shore on 26 November, and one on M1 on 23 November 1997 were the only records until a flock of up to eight appeared on the shore near the end of Hesketh Road on

29 November 1998, remaining until 26 February 1999. The most recent record is of one on the peninsula on 17 March 1999.

ORTOLAN BUNTING *Emberiza hortulana*

A scarce migrant at east and south coast sites, the Ortolan is a very rare vagrant to North-West England. One on 31 October 1973 is the only record for Marshside.

YELLOWHAMMER *Emberiza citrinella*

Up to the mid-1980s Yellowhammers bred quite commonly in South-West Lancashire and appreciable winter flocks were regularly seen at favoured mossland and coastal sites. There seems to have been some decline in the population in the last decade. At Marshside Yellowhammers appear always to have been fairly scarce and irregular visitors, mainly in late winter and spring. Two on 29 May 1989, single birds on 1 and 4 April 1992, 26 March 1994 and 14 September 1995, two on 7 January and singles on 24 March 1996, 6 and 16 October 1998 and in September 1999 are the only documented occurrences during the last 15 years.

REED BUNTING *Emberiza schoeniclus*

The Reed Bunting, with its cheerful if unmusical song delivered from the top of a reed stem or willow is one of Marshside's most familiar passerines, although nesting numbers have decreased substantially since the mid-1980s. An estimated total of 20 pairs bred along the foot of the Marine Drive embankment, in the SSSI ditch and on the inner portion of the saltmarsh in 1984, and at least 18 pairs in the following year. By the 1998 and 1999 breeding seasons this total had declined to ten pairs, at the most.

Reed Buntings also occur as passage migrants in late autumn and early spring. Numbers are never spectacular but 40 were seen on 13 October 1992 and 45 on 29 October 1995, while 35 on 25 February 1995 is the highest spring total recorded during the 1990s. A variable and apparently declining population of Reed Buntings overwinters, mainly on the saltmarsh; over 80 were counted on Crossens saltmarsh alone in January 1984 but totals for the whole area in the 1990s have seldom exceeded 50 birds, and numbers barely reached double figures in the 1996-97 and subsequent winters.

CORN BUNTING *Miliaria calandra*

The Corn Bunting has decreased quite catastrophically across most of Britain and Ireland since the early 1970s (Gibbons, Reid & Chapman, 1993). The population on the South-West Lancashire mosslands appears to have fared rather better than most; nevertheless, some contraction seems to have occurred even here and the species is now seldom seen on the reclaimed marsh at Banks, a former stronghold. At Marshside-Crossens the decline in the Corn Bunting's fortunes has been dramatic.

As recently as the late 1970s and early 1980s moderate-sized flocks, presumably from the Banks marsh population, were frequent visitors, particularly in winter and early spring. 40-plus on 22 January 1977, 25 by the Sand Plant on 12 March 1980 and 65 along the Marine Drive embankment north of the Wildfowlers' car park on 29 January 1984 are the highest of many counts extracted from field notebooks of the period. A few pairs nested on Crossens saltmarsh in most years up to 1984, but not thereafter.

Records at any time of year simply dwindled away after 1985, and a total of five birds in spring 1989, singles in January 1991 and March 1993, four on the saltmarsh on 6 and 7 February 1996, and one on M1 on 14 May 1999 are the only documented occurrences in the last decade.

References

Cranswick, P. A., Waters, R. J., Evans, J. & Pollitt, M. S. (Eds)
The Wetland Bird Survey 1993-94: Wildfowl and Wader Counts
BTO/WWT/RSPB Joint Nature Conservation Committee 1995

Gibbons, D. W., Reid, J. B. & Chapman, R. A. (Eds)
The New Atlas of Breeding Birds in Britain and Ireland: 1988-1991
London: T & A. D. Poyser 1993

Greenhalgh, M. E.
Wildfowl of the Ribble Estuary
Chester: WAGBI 1975

Jones, M. (Ed)
1987 Lancashire Bird Report
Lancashire & Cheshire Fauna Society 1988

Jones, M. (Ed)
1994 Lancashire Bird Report
Lancashire and Cheshire Fauna Society 1995

Lack, P. (Ed)
The Atlas of Wintering Birds in Britain and Ireland
Calton: T. & A. D. Poyser 1986

Oakes, C.
The Birds of Lancashire
Edinburgh & London: Oliver & Boyd 1953

Ogilvie, M.A.
Wild Geese
Berkhamsted: T. & A. D. Poyser 1978

Spencer, K.G.
The Status and Distribution of Birds in Lancashire
Burnley: 1973

Spencer, K. G.
1975 Lancashire Bird Report
Lancashire and Cheshire Fauna Society 1976

the fly p
and other stories

*Memories are like fish in the evening.
Only the ripples show.*

The Fly Pool / John McAllister -- 1st ed.
Published in the UK in 2003 by Black Mountain Press
PO Box 9, Ballyclare, Northern Ireland BT39 0JW

Copyright © John McAllister 2003

AWARDS FOR ALL

The publishers wish to thank Awards for All Northern Ireland for financially supporting this publication.

ISBN 0-9537570-4-8

This book is a work of fiction. Name, characters, places and incidents either are products of the author's imagination or are used fictitiously. Any resemblance to actual events or locales or persons, living or dead, is entirely coincidental.

The moral right of the author has been asserted. A CIP catalogue record for this book is available from the British Library. No part of this publication may be transmitted in any form or by any means, electronic or mechanical, including photography, recording or any information storage or retrieval system, without permission in writing from the publisher. The book is sold subject to the condition that it shall not, by way of trade or otherwise, be lent, resold or otherwise circulated without the publisher's prior consent in any form of binding or cover other than that in which it is published and without a similar condition, including this condition, being imposed on the subsequent publisher.

Cover design/Typeset: 13AD design
Cover photograph: Darren Brown
Author photograph: © Michael Campbell, Armagh
Printer: GPS Colour Graphics

the fly pool
and other stories

John McAllister

The Black Mountain Press, Ballyclare

Acknowledgements

Stories have appeared or are due to appear in the following publications: *A Room Full of Nothing but our Lives*, *A Stroll Through the Fields*, *Alchemy*, *The Black Mountain Review*, *Breaking the Skin Anthology*, *Northern Woman*, *Shots*, *The London Magazine*.
In addition, some stories were finalists in the Allingham, Martin Healy and Cootehill Literary Competitions, or were broadcast on Armagh's *Talking Newspaper* and on Dublin South Radio.

When your writing days are as long in the tooth as mine are, you have an incredible number of people to thank for a diversity of reasons: the members of the Omagh Writers Group; the Armagh Writers Group; Writers in Residence, Carol Rumens, Colin Teevan, Daragh Carville, Sinéad Morrissey and the members of the Queens Writers Group; Professors Gerald Dawe, Brendan Kennelly, Tom Kilroy and Jonathan Williams, Writer-in-Residence Marion Carr and fellow members of the 1998-99 M.Phil. course in Creative Writing, TCD; my wife, Trish, and children, Lucie and Daniel, and my favourite mother-in-law, Bernadette Byrne; my sisters, Big Mary and Our Pat, and their families and, especially, my nieces, Sarah Mallon and her husband Garett, Barbara Campbell and her partner Brian Maginness, for putting me up during my time at TCD; Sam Burnside of the Derry Verbal Arts Centre; Lois Bennett, Roy Cummings, Carol and Stephen Day, Thomas Foulks, Margaret and Arthur Gibson, Kevin and Patricia Gormley, Kevin Hart, Des Kenny, Nigel McLoughlin and my agent Mike Shaw. A special word of thanks to novelist, Glenn Patterson, for a decade of encouragement, and to my first mentor, Damian Quinn (1961 – 2001, RIP).

contents

1. THE CASE BOOK OF SERGEANT BARLOW

CURLES BRIDGE	9
TAKING STOCK	16
MAKING GOOD	25
PROOF	34
HANG-UP	44

2. PLAYING WITH FIRE

CICADA	55
PLAYING WITH FIRE	68
MADONNA OF THE FALL	72
THE PRIDE OF HENRY GEORGE	86
WAITING TO GO ON	95
THE FLY POOL	100
GROWING PAINS	107
DOUGHNUTS	114
THIS LIFE	117
SUMMERLAND	142
WILFULLY AND DELIBERATELY	148
SCORING	161

to Trish

Red jumper and jeans curled up in my seat
Near the bookcase lamp and coal fire heat.
Mind buried in book, specs from nose dripping
Winter has come, she is hibernating.

THE CASE BOOK OF SERGEANT BARLOW

CURLES BRIDGE

Police Sergeant Barlow looked at the portrait of the Queen. He, his old boss, and the King had taken up their appointments on the same day. Now the other two were gone, and Acting District Inspector Harvey had the King's portrait down and Elizabeth's up within an hour of taking over. Not that Barlow had anything against the new Queen.

Barlow stood on, oblivious to the altercation at the front desk between Constable Jackson and the tramp. Jackson's temper was beginning to unravel. 'This isn't a hotel, you know. And leave that fire alone.'

Mr Edward Adair, gentleman, and tramp by profession, gave the coals a final poke. 'But it's Christmas, no man should be alone at Christmas. Please inform Mr Barlow that Edward would appreciate the same room.'

'Cells. They're cells. And you can't just book in.'

'But, my dear chap, I always do.'

Jackson leaned over the counter and threatened Edward with a pencil. 'If you don't leave right now I'll do you for wasting police time.'

Edward smiled and warmed his hands at the flames. Jackson sighed in relief as Barlow paced slowly into the room, his face set in a pout of thought. Jackson said, 'Sergeant, this man won't go.'

'Changed times, Edward,' said Barlow, to the tramp.

'And my accommodation for the festive season?'

'New inspector, new rules. There's more than you catching it.'

Edward vibrated with indignation. 'So it's like that, Mr Barlow. After all these years?'

'Aye.'

'And a happy Christmas to you,' said Edward stiffly, and left.

Jackson scratched at himself to remove the fleas. 'Sarge, who is that old drunk?'

Barlow shot behind the desk and prodded Jackson hard in the chest. 'That old drunk, as you call him, is *Mister* Edward Adair.'

He prodded a second time. Jackson said, 'Easy, Sarge,' and retreated.

'And *Mister* Edward Adair is a friend of the Chairman of the Police Authority. And....' The finger pushed gently against Jackson's cringing breastbone, 'Call me "Sarge" again, and you'll be on nights between now and kingdom come.'

Jackson heard Acting District Inspector Harvey's tiptoe step in the corridor; he made himself busy straightening the Incident Book.

Barlow raised his voice. 'Are you looking for something, sir?'

Harvey stepped into view. 'Jackson, new Orders of the Day.' He slapped a sheet of paper on the counter and marched off.

The sheet said:

DAILY ORDERS

Kirktown is now a 'no crime' area. Every misdemeanour will be noted with a view to prosecution.

Barlow grunted and left, he was due on duty again that night for the Annual Rotary Ball. By tradition it was held in the town hall on the Friday before Christmas, and Sergeant Barlow was always there to greet the Rotarians and their guests as they arrived. Early as he was the tramp, Edward, was there before him.

'It's cold,' said Edward, and pulled his greatcoat tighter round him.

'No it's not.'

'Whatever pleases you.' Edward retreated to the other side of the steps.

Two of the Dunlops from Hell's Kitchen passed. They were taking an interest in the parked cars.

Constable Jackson stood nearby, trying to look useful. Barlow called him over. 'Move that lot on, son.'

Jackson squared himself off and set his hat firmly on his head. Barlow said, 'Those boys have a quick temper. You start anything, and you're on your own.'

He left Jackson to it and turned to greet the mayor, Sam Martin, and his good lady as they arrived in the official limousine. The mayor acknowledged Barlow's salute with a grunt and gave Edward a ten-shilling note.

'God bless you, and a merry Christmas,' said Edward.

Barlow looked on sourly as the note disappeared into a pocket. 'You wouldn't like to disappear as well?' Another couple came and again a ten-shilling note changed hands.

Acting District Inspector Harvey drove up in his new Triumph saloon. He wound down the window. 'Barlow, what are you doing here?'

'Tradition, sir. The Station Sergeant sees the Rotarians into their Christmas do.'

Harvey bristled. 'And makes sure of his Christmas tips in the process. That is one tradition I intend to stop.'

Instead of replying, Barlow walked round the car and opened the passenger door for Harvey's companion. 'Have a nice evening, Mrs Harvey.'

Harvey said, 'This is Mrs Carberry, a cousin of our Chairman. Her husband is away on business...' He stopped. 'Really, Barlow, it's got nothing to do with you.'

'No, sir. Of course not, sir.' Barlow nodded, he all but winked.

Harvey turned stiffly away, and saw Edward pocketing another ten-shilling note. 'Sergeant, what is that *thing* doing here?'

'Making a fortune,' said Barlow.

'Good evening, Edward,' said Mrs Carberry, and gave him a pound note.

She walked on. Harvey had to follow, choking.

'Clear off,' Barlow told Edward.

'But my dear chap, this is my best night.'

'Take it up with the new Station Sergeant. I'm being posted.'

'After all these years?' Edward tut-tutted to himself and wandered off in the direction of the Bridge Bar.

Barlow came in early the next morning, and found Constable Jackson draped over the counter. 'What's up with you?' Barlow asked. He sniffed and got the smell of cigarettes and of the great unwashed.

Jackson levered himself onto his elbows. 'Not a wink of sleep last night. I can't be doing with it, Sergeant.'

'You young boys, you've got no stamina.'

Barlow reached for the Incident Book. Two names leaped out at him. Both Dunlops, both charged with robbing the Kirktown Brewery. 'What buck eejit…?' He bit off his words.

Jackson blinked rapidly to keep his eyes open. 'The Dunlop women were in half the night, drunk as lords and swearing like Tam the devil; we thought they would never go. Their kids were everywhere. My pencil was nicked three times.'

'Bollocks,' said Barlow, as Edward appeared, beaming from ear to ear.

'Mr Barlow, I want to place myself in police custody.'

'Not now, Edward, not now.'

Edward approached Jackson. 'Young man,' he put his hand on his heart and intoned the words as if swearing a great oath, 'I, Edward Charles George Adair, confess that I did, wilfully and deliberately, and with malice aforethought, cause or compel, by means of threats or other illegal means, the payment to me of monies by way of blackmail or extortion. And that in addition…'

Jackson's mouth was opening wider and wider. Barlow grabbed Edward by the collar and dragged him to the fire. 'What the hell are you talking about?'

'Blackmail. Blackmail and being a Peeping Tom.'

'Are you finally loony?'

Edward smiled. 'As you are aware, my habitual abode is situated under Curles Bridge. After each ball, a number of the participants retire to the adjoins and surrounds of the bridge, for social intercourse of a highly intimate nature.'

'You mean shagging? You dirty old man.'

'Shagging yes, dirty no.' Edward looked offended.

'So you watch?'

'One ascertains, one identifies. A nod here, a wink there, everything perfectly civil, and my comfort is secured, my edification enhanced.' He pulled a piece of paper from his pocket and handed it over. 'Proof of my culpability. Last night's participants.'

Barlow said, 'You don't want to be here this year, we've got the Dunlops in.'

'Very nice gentlemen. They supply me with a bottle of best Scotch every festive season.'

'Would you shut your mouth,' said Barlow, and looked at the list of cars. It gave their make and registration numbers.

Edward said, 'Some of the participants, particularly the ladies, tend to be exhibitionist by nature.'

'What's exhibitionist?'

'They find that an appreciative audience enhances their enjoyment.'

'A year,' said Barlow. 'A year of a beer and chaser every Saturday night.' He roared for Jackson as if he was at the other end of the building instead of steps away. 'Book him?'

Jackson grabbed a pencil. 'What charge?'

'Annoying my bloody head!' snarled Barlow.

'I know the way,' said Edward, and disappeared down the corridor. He was back in a moment. 'Mr Jackson, how do you like your tea? And would you find an Ulster fry acceptable if the ingredients happen to be in situ?'

Jackson didn't know what to say. The phone rang, he grabbed it in relief. 'Yes, sir. Right away, sir.' He said to Barlow. 'Inspector Harvey's looking for you.'

'Tell him to wait,' said Barlow, and copied the details from Edward's list into his notebook. He wrote slowly, and leaned heavily on the pencil to make sure the registrations numbers showed up clear. Then he went off to Harvey's office where he found Harvey in a good mood.

The edges of Harvey's lips softened, which was the nearest he ever got to a smile. 'Well, Barlow, what do you think?'

'Sir?'

'The Dunlops. They robbed the brewery every Christmas, and always on the night of the Rotary Ball. That is what I have against you, Barlow, you haven't the intelligence to see the obvious.'

'And you caught the Dunlops, sir?'

'Red handed.'

Harvey waited for Barlow to congratulate him. It didn't come. He flushed. 'Clear your things, I don't want you back in this station.'

Barlow nodded and wandered to the door. He stopped with his hand on the handle. 'I was wondering, sir. The indecency charges, you'll want me back to testify?'

Harvey looked up from his paperwork. 'What indecency charges?'

Barlow became enthusiastic. 'This no crime thing of yours, sir. You're right, we had got slack and complacent.'

'What indecency charges?'

'The cars around Curles Bridge last night.' Barlow pulled out his notebook and thumbed slowly through the pages. 'About half a dozen, sir, and one was a spank new Triumph just like yours.'

'What?'

Barlow handed the book over to Harvey. 'I've no head for numbers, sir, that's why I wrote them down, careful like.' He pointed to one car on the list. 'But I can swear both those parties were totally not dressed.'

Harvey was sweating; Barlow didn't appear to notice.

The Chairman of the Police Authority burst in, red-faced with fury. 'Harvey, you're an idiot. Arresting the Dunlops? What the hell were you thinking of?'

Harvey tiptoed swiftly round the desk to greet the Chairman.

'They were robbing...'

'They do that every Christmas, dammit!'

'They what?' Harvey went grey around the jawbone.

'Every Christmas,' roared the Chairman. 'And while you were charging them, the rest of the family broke in and stole a lorry load of drink.'

Harvey's grey started to look bilious.

The Chairman almost danced with rage. 'It was an understanding. They get what does them for Christmas and they leave me alone for the rest of the year. And they make sure all the other hoods do the same.' He jabbed his face hard against Harvey's. 'Don't forget, Harvey, you're only here on trial. Another cock-up like this....' The Chairman stormed out.

Barlow's notebook was starting to crumble in Harvey's hands. It took a great effort for him to relax his grip. He went back and sat at his desk. 'Sergeant... I....' He didn't seem able to get enough air. He played with his pen; he put it down and picked it up again.

'The lady was exhibitionist,' said Barlow.

Harvey wiped sweat from his forehead. He gritted his teeth and tried to sound pleasant. 'We've had a bad start. Perhaps when we get to know each other better?' He took a notebook out of a drawer and slid it over. 'A new notebook, a new year, a fresh beginning?'

'Fine by me, sir,' said Barlow.

TAKING STOCK

Along the river there were only farm dogs and evasive answers. They were the least of Sergeant Barlow's problems, for the rain wasn't taking time to come down and the Dunlops were playing their usual Orange card. For most of the year it was merely a case of injuries sustained while resisting arrest, but at the beginning of July, with the marching season in full swing, and them good Orangemen and Barlow a Catholic, a *Roman* Catholic, then it was police brutality.

A sheepdog shot out of an old shed, all hackles and bared teeth, barking. Barlow took his foot off the pedal and kicked at it. The dog swerved and the bicycle wobbled dangerously. He kicked again, and turned into the lane leading to the Widow Todd's house. Across from the entrance to the lane was the farm belonging to the Chairman of the Police Authority. Not that he lived there of course, not grand enough for the wife. District Inspector Harvey would personally visit the Chairman at his town house, drink a whisky and record a pack of lies. 'My dear, Harvey, me plant a crop of beet?'

'Lying git,' said Barlow into the rain.

Barlow dumped the bike against the farmhouse wall. The half-door leading into a small porch was open. The porch was dry because the rain was driving away from it. He banged the upper door. When the widow appeared he held up a ledger wrapped in oilcloth. 'Census, missis.'

She nodded, it was a nervous habit she had, of her body agreeing with her mind before she said or did anything. 'Come in.' There was no warmth in her voice. She led the way into the kitchen and put the kettle on the range.

He stood uncertain, his cape already forming puddles on the flagged floor. 'Your crops, the acreage... the usual.'

She pushed a chair away from the table. 'Another year.'

'And a bad one too, missis.' The deal table was scrubbed white; his hat made a dark stain of damp. He took it off again. 'But it's near the Twelfth, and they say God's an Orangeman.'

Barlow hung his hat and cape out in the porch, then he unwrapped the ledger and put the oilcloth on the floor. He was relieved to see that the damp stain on the table was already fading.

He worked ponderously through the ledger. He could have answered the questions himself because the widow cycled the same crops: wheat, root and fallow year by year. Only the bog meadow along the wandering river remained forever in grass. He kept his eye on the page because each answer, whether yes or no, was accompanied by a nod, and he found that confusing.

A mug of tea sweetened with two spoonfuls of sugar appeared in front of him. He drained it and it was promptly refilled. In between times, she stirred a pot of stew for her supper and an even bigger one of porridge for the pigs in the morning. Barlow was so hungry he could have eaten the porridge, but there was no offer of food.

The widow interrupted his questions. 'There must be more to life than your ledger.'

He looked up, startled, and saw an angry pain on her face. He said, 'Sometimes I wonder myself, missis.'

He wasn't sure what else was expected of him so he drank his tea. The widow nodded vigorously and turned back to her stirring. Once the ledger was filled in, Barlow swung the cape over his shoulders and left.

'Funny woman,' he said to himself as the bike creaked down the lane. It was always the same with her, never welcoming, but never unfriendly. A kind soul, people said, with her only son dead in the war, and her husband before that. With fifty acres she could have got another husband, but every approach had been met with a steady rebuff.

At the bottom of the lane the dog came at him again. He was ready for it this time and his boot nearly connected. The dog gave a yelp and disappeared into the shed. Barlow felt better as he pedalled the long miles back to the station.

District Inspector Harvey and the Chairman of the Police Authority were waiting for Barlow. Barlow didn't mind his clothes dripping onto

Harvey's carpet.

Harvey sat at his desk. The Chairman stood to the right of the Queen's picture: upright, firm, hands behind back. Crisp as any soldier on parade, though he had never been more than a quartermaster, serving in the Home Counties throughout the whole of the war.

Harvey said that he and the Chairman had been talking, the local Twelfth parade needed careful handling. It wasn't just a question of no triumphalism while marching through the Catholic area of the town to the railway station. There was also the point that he, Barlow, led the parade. 'No harm in that,' said Harvey, and hastily added he was sure that Barlow's loyalty was to his uniform and his fellow officers.

Barlow nodded. If he had a *political* opinion it had to do with Chairmen who were never out of the station.

The Chairman said, 'We are particularly concerned this year because of the complaints lodged against you by members of the Dunlop family.'

He looked annoyed, something he did anytime the Dunlops were mentioned; the lorry load of drink stolen from his brewery at Christmas still rankled. 'The Dunlops, as you well know, are stalwarts of the Lodge, they carry the banner. So it would be unseemly if…' He looked over at Harvey. 'Laurence, you explain.'

The explanation was simple. Harvey had deployed his forces to make sure there would be no trouble. A ring of policemen would accompany the parade to the railway station.

Barlow said, 'Last year there was only me, and a constable bringing up the rear.'

Harvey flushed. 'Are you questioning my judgment?'

The Chairman's words boomed in the room, he had that sort of voice. 'Barlow, I have nothing against you leading the parade. But what you do at the Catholic church, really!'

Barlow looked at the Queen and wondered what her opinion was. 'It's like this, sir. A lot of the Orangemen are old soldiers like me.' His eyes drifted to where Harvey would have worn his medal ribbons, if he had any. 'They like a bit of swank so I call the time. And when you're on parade you have to salute somebody.'

Harvey was outraged. 'An *eyes right* at the Catholic church?' The Chairman coughed and Harvey controlled his temper. He was determined, he said, that it wouldn't happen again. And, anyway, Barlow had a more important job to do this year. There had been a lot of cattle stealing in the area recently, and the Chairman was concerned that somebody would try to steal his prize Herefords, particularly on the Twelfth when he and his men were away with the Lodge. Harvey's instructions were detailed and took nearly five minutes to deliver.

Finally, he asked, 'Any questions?'

Barlow finished writing into his notebook, *from eight at night to six in the morning, and all day the Twelfth*. He put his pencil away, he put his notebook away.

He stood in thought. 'Sir, seeing you brought the subject up. Why does the parade go past the Catholic church when the train station is at the other end of town?'

The air crackled with indignation. Barlow saluted the Queen and left.

The shed was open-sided. The dog lay in the only dry corner, eyeing Barlow warily. The Herefords were in the field across the way or, rather, Barlow hoped they were. When the light went out of the day they were sheltering against the far hedge, and he had no intention of going looking for them. He was wet already, from his feet to well above the hang of his cape, and from the collar down. The two damps met at the seat of his pants.

A car came crawling along the road, its lights picking out the shafts of rain angling into Barlow's face. He muttered to himself and stepped out when the car stopped beside him. The dog barked, Barlow told it to shut up.

It was District Inspector Harvey checking up on him. 'Nothing to report, sir,' he said.

'Stay alert, I might be back later.'

Barlow watched Harvey drive off. He looked at the dog; it was back in its bed. 'You could have piddled on its wheels.'

The widow appeared. She was wearing a man's old coat thrown over her clothes. 'Come up to the house.'

He was glad to follow her, summer or not there were blocks of ice where his feet should have been. The kitchen was warm and she had supper on the table: the remains of the stew, and homemade wheaten bread. The kettle was already singing on the stove. She pushed a chair out for him and he dared loosen his jacket by a button.

She nodded. 'Take it off.' He did, and she hung it before the range while he ladled stew onto the wheaten bread and chomped at it. His jacket started to steam in the heat. She nodded again. 'Boots, too.'

He gulped his mouthful of food down; he begrudged wasting the taste and the heat. 'I'm supposed to be on guard duty, missis.'

'The dog will bark if anyone comes.'

She had been cleaning the family silver. One picture in a silver frame was of her and her husband on their wedding day. She standing taller than him, he always was a weed of a man, and her looking out of place in white.

He said, 'You were a fine looking woman.'

She said, 'Were.' The photograph was wiped and put away. 'All these years, and nothing to show for it but hard work.' She fussed with the kettle at the range, nodding all the while. 'I come from a big family. I never slept on my own until I was a widow.'

Barlow ate slowly rather than commit himself with words. She put his drink down before him, a hot whiskey with cloves, a strong one. She joined him with her own whiskey and they drank them in silence.

'Smoke, I know you do,' she said. He lit a Woodbine and puffed slowly while she cleared away. He offered her one and she said, 'No.' He kept the pack held out. She hesitated, nodded and took one, nodded again and drew the smoke into her lungs. She let it out with a sigh of pleasure.

When it was finished she stood over him. 'I'll go up.' She pointed to the ceiling above his head. 'Keep the window open and you'll hear the dog.'

He had another Woodbine and watched its swirls of smoke along the ceiling, and listened to her footsteps as she got ready for bed. When the house was silent again, he stripped off his clothes and hung them before the range to dry. Finally he was down to his combinations, one-piece underwear that stretched from knee to wrist. Their whiteness had faded with age. After a bit of thought, Barlow removed them and went up the stairs to the bedroom above the kitchen.

He thought it might be her room. It was delicate, all little fancies and full of smells. He was sure when he got into the double bed, it had a hot water bottle already going cold. The widow was there as well. She was naked and waiting.

The singing of the birds woke Barlow. The early daylight glaring in the open window had made him restless. He looked at the widow. She was lying with her hands behind her head, staring at the ceiling.

'It's the Twelfth,' she said.

'Of course it is, missis, isn't the sun shining?'

Barlow was nearly glad of it after three nights with little sleep. He looked at her, hoping. The widow liked to make the first move.

She kept staring at the ceiling. 'I'm due another call about pulling the Ragweed.'

'You are that.' He risked a hand on her leg.

She rose over him, the blankets falling away from her breasts; they sagged on her bony frame. 'When the time comes, Mr Barlow, would you send another constable?'

'Now why would I be doing that?' He turned her under him.

The dog barked, he tried to ignore it. It barked again.

He got out of bed and looked out the window, afraid it was Harvey making an early morning check. He could see some sort of vehicle parked near the bottom of the lane. 'I'll be back.'

She nodded her head then she made it shake. 'It must stop here, Mr Barlow. My neighbours wouldn't like it, not with a Catholic.'

'Nor my wife.'

He said it deliberately.

Her nod was slow and thoughtful. 'I was aware of that too.' There were tears in the widow's eyes as she held out her hand.

Barlow nearly spat before they shook as if settling a deal. He dressed quickly and left the house.

He was at the end of the lane before the echoing thud of cattle hooves on wood seeped through his abstraction. 'Begod!' he said, and drew his truncheon.

The dog came out to meet him. 'Git,' he ordered. The dog ignored the order and stayed quietly at his heels as he peeked up the county road. A lorry was backed into the gateway of the Chairman's field. He could see the shapes of three men driving cattle into it with sticks and whispered shouts. Two animals were already in the lorry. He tiptoed down the road, keeping the bulk of the lorry between himself and the men. They were busy loading a third animal.

One of the men asked. 'What's the difference between Sergeant Barlow and one of these cows?'

Barlow smiled. 'Begod, it's the Dunlops.'

The man answered his own question. 'The cow's eyes are wider apart.'

There was a roar of laughter.

Barlow stepped around the side of the lorry. He swung his truncheon and caught the comedian a crack on the point of the shoulder. The second man was still laughing when Barlow sideswiped his elbow. The third one saw him coming, and had the advantage of being at the top of the loading ramp. He swung his boot at Barlow's head. Barlow caught the man's knee with the truncheon and he fell in among the cattle.

It was three to one against so Barlow went over the Dunlops twice more: elbow, knee and shoulder, to make sure they were properly subdued. Then he prodded them with the truncheon, and the dog nipped at their ankles, until they were in the lorry.

Barlow swung the ramp up and secured it, leaving them among the cattle. He tried to pat the dog, but it snarled at him so he swung his boot. It seemed happier as it retreated into the shed.

District Inspector Harvey was all bustle and fuss. The parade had started from the Orange Hall late because they had waited for their banner men to turn up. Finally, the Chairman had dammed the Dunlops to hell and given the order to march.

Harvey had gone ahead to the Catholic church. Edward Adair, gentleman, and tramp of the parish was in attendance. His old greatcoat had been brushed down, and he was wearing a frayed suit and tie. He nodded politely to Harvey. Other than Harvey and Edward, the whole area was deserted except for some children swinging off the railings, and two policemen for each child. Harvey began to think he could have used less security to push the parade through.

He tensed as the off-notes played by the Kirktown Silver Band died away. The morning air was filled with the sound of a single kettledrum and the tramp of marching feet. He checked again for trouble as the Lodge and its banner swung into view round the corner.

An old lorry came spluttering along the road. A policeman stepped out and held up his hand. It drove on, and he had to jump clear as its wheels threatened to nip his toes. It stopped at the Catholic church and Barlow got out. He stretched and yawned.

Harvey came storming up. 'Barlow, what are you doing here?'

'Bringing your banner men, sir.'

Barlow went round to the back of the lorry and let down the ramp. The Dunlops hobbled out, they were covered in muck. The cattle tried to follow them; Barlow drove the animals back with a roar.

The three men huddled together as the policemen closed in on them. There was a lot of laughter and the holding of noses. Nobody went too close or got too funny; the Dunlops were always dangerous and had long memories.

Harvey quickly questioned Barlow about the attempted theft of the cattle. He ordered the Dunlops to be led away to the cells. Their bruised muscles had stiffened and they could only move slowly.

Harvey was outraged. 'These men are injured.'

Edward looked at District Inspector Harvey over a raised nose. A thousand years of Adair nobility went into the cold stare. 'Three against one, a prima facie case of police brutality.'

A shout of laughter went up from the watching policemen.

The Orange parade was almost on top of them; Harvey wanted the lorry moved. The engine churned, but would not kick into life. He rushed off to try the starting-handle himself.

Barlow looked in disgust at the marchers who could see something going on at the lorry. As old soldiers they knew to keep looking ahead, but they had lost their concentration and the rhythm of their stride was beginning to break up. The Chairman was at the front of the parade carrying the silver sword. His face turned beetroot red when he recognised the Dunlops and his cattle.

Barlow began to call the time. 'Left... Left...' The marchers stride

picked up, their arms swung that bit higher.
　　When they drew level with the Catholic church, Barlow roared. 'Parade. Eyes... right!'

MAKING GOOD

Barlow found Edward slumped over generations of drink stains. He joined him and, with a slow deliberate gesture, placed his cap on the bar counter. It was eight thirty on a Saturday evening, and Sergeant Barlow of the Royal Ulster Constabulary was off duty.

'A pint and a chaser for Mr Edward,' ordered Barlow, over the heads of men waiting to be served.

'That is extremely gracious of you,' said Edward. His hand was out for the whiskey before the barman could start his pour.

'It's my shout,' said Barlow.

He paid for the order with a half-crown, accepted his change and a pint of Guinness on the house, and watched Edward in the back mirror as he did so. Edward had come down a long way from the man who used to smile at him over unexploded bombs. Drink was just as lethal, but a lot slower.

For all Edward's rush for the whiskey he hardly touched it. Barlow asked, 'What devil of a memory is eating at you now?'

'Nothing.'

'When you say "nothing"...'

'I mean, nothing!' Edward drained the whiskey quickly. 'Corporal Anderson...'

Barlow snorted. 'Anderson died of Consumption after the war.'

Edward indicated the door to the *Select*. 'Corporal Anderson's son, young Robert, awaits your pleasure in a matter pertaining to your professional competence.'

Barlow's brow furrowed in thought as he sucked Guinness down his throat. Edward drank his pint and made a play of not having enough coins for a refill.

'Pertaining?' asked Barlow.

'Pertaining.'

Young Robert Anderson was the sole occupant of the *Select*. The boy was thin and bony, and sat in the middle of the room where he could keep a wary eye on the inner and outer doors.

He remained seated when Barlow appeared. 'My mother sent me.' He pushed Barlow's Club Orange refill away. 'It's a breach of fiduciary trust, I could go to jail for six months.'

'So you nicked some money then?'

Robby's face blazed. 'I never...'

Barlow thumped the table. 'Stop bothering me with excuses on a Saturday night.'

Robby stormed to the door. Barlow stepped up behind him. He was surprised at the strength of the boy's pull on the handle, and had to exert force to keep the door shut.

Robby said, into the door. 'I'm entitled to the solicitor of my choice.'

'Quite the little lawyer, aren't we?'

Robby looked round, his eyes doubtful. Barlow led him back to the table. 'You drink that squash or I'll give you a thick ear, see if I don't.'

Mrs Anderson had hopes of getting Robby a post as a solicitor's clerk. In the meantime, he had taken a job with the Kirktown Investment Company where he ran the messages and helped out as required. When a Mrs Kyle rang in a panic that her husband had gone off without that day's lodgement, the manager had sent Robby to pick it up.

It was a long cycle out to Kyle's house on a warm day, and Robby was glad to be invited in for a drink. His mouth tightened when he spoke of the kitchen. Where there was wall there was embroidered tracts. *Put your trust in the Lord* was the one he remembered.

Mrs Kyle seemed to be glad of Robbie's company and wanted him to stay for lunch, but he said it was too early and the manager would be wondering where he had got to. All the time the money sat on the kitchen table in front of him; it was in a brown paper package, tied the way butchers did with the string pulled until the contents were squeezed. He tried undoing the knots, but they were too tight and he didn't like to insist on counting the money first, the manager hadn't told him to, so he took the package and went straight back to the office.

'No stopping off anywhere?' queried Barlow.

The boy said, 'Her and her two-faced religion. The money was never there.'

Once Robby reached the office the string was cut and the money counted. The manager rang Mrs Kyle to confirm the lodgement at five hundred and forty-five pounds, she was adamant that there should be six hundred in cash. The manager called an emergency meeting of the directors, and the senior director available questioned Robby in the boardroom.

'Who would that be?' asked Barlow.

'Mr Moncrief.'

'Solicitor Moncrief?'

There was something in the tone of Barlow's voice that brought Robby briefly out of his sulk of temper. 'Yes.'

'What did Solicitor Moncrief say?'

The boy was playing with his drink; it gave his hands something to do. 'He said it didn't matter whether I stole the money or not, I was responsible for the shortfall and the loss had to be made good.'

'Aye,' growled Barlow. 'That would be just like him.' He pulled a packet of Players from his pocket. 'Can the money be got?'

'It already has.' Robby flinched from the doubt in Barlow's face; orange squash jibbed over the table. 'My father's train set, we had to sell it.' He started to cry.

Barlow took his time about lighting a cigarette. Robert Senior's train set was something special, the hand-carved figures were done to scale, and painted so delicately you felt you could seen the individual hairs on the head of the people. The train set was the talk of the countryside.

Barlow told Robby to blow his nose then he said, 'You must have got a fair bit for it.'

Robby scrunched the handkerchief until his knuckles went white. 'Fifty-five pounds.'

Barlow started. 'It was worth more than that.'

'That's all Solicitor Moncrief's son would give me.' Robby's eyes were full of tears.

They weren't his father's eyes, Barlow thought, though they could have been his mother's. 'Young Moncrief?' he said, and drew so slowly on his cigarette he started to think about going back on the pipe. He had a thoughtful final pint while writing the names of the people involved into his notebook: the manager and staff of the Investment Company, Solicitor Moncrief, Moncrief's son, Mr and Mrs Kyle. The whole thing nagged at him over the weekend, and there was something about the sum of fifty-five pounds that rang a bell. He thought he would start by having a word with Mrs Kyle.

Mrs Kyle was startled to see Barlow on the backdoor step. The movement of hand to heart was the Presbyterian equivalent of blessing herself to ward off the evil eye.

'Just a word, missis,' he said, and politely bullied his way past the scrawny little woman into the house. He looked at the kitchen units with their Formica finish in royal blue, instead of the oilcloth and painted wood he was used to. 'Very nice,' he said. He thought the steel sink didn't give a man enough elbowroom for a good wash.

'The Lord has been good to us,' she intoned, rushing to tidy away her reading glasses and yesterday's paper.

Barlow was made sit at the table while the kettle was put on to boil. Mrs Kyle ladened the table with an array of cold cuts and sweet things, and still worried about what she had forgotten. She sat across from Barlow with a cup of tea in her hands.

'Very nice,' he said again, looking from her to the tracts on the wall. One said *Judge not lest ye be judged*.

She put her cup down, it rattled firmly into the saucer. 'I'm not going to say one word against Robby Anderson. Wilson says it's all my fault for panicking, he would have got the money lodged no matter what was happening at the factory.'

'He's a busy man then?'

She sat prim with her hands cupped in her lap. 'He works hard. I can hardly begrudge him his nights out with his friends. *A little wine is good for the stomach*.' She looked achingly lonely.

Barlow nodded – anytime he saw Wilson Kyle drinking it was double brandies. He lathered a slice of soda with butter and chewed thoughtfully. 'Lovely, missis, you're a great cook.' She smiled, pleased.

He cleared his mouth with a drink of tea. 'This missing money?'

She was uneasy under Barlow's steady stare, and got up and tidied things into the sink. 'I'm not sure now if there was ever six hundred pounds in the bag in the first place.'

Barlow was angry with Edward for getting him involved. There was no case to investigate unless somebody was willing to complain, and Mrs Kyle clearly wasn't. He was equally angry with Robby for getting into such a mess, he was the one kid he always thought would make good.

'So what happened about the money and the counting?'

Mrs Kyle ran water hard as if to bury her words, steam curled across the room. 'Miss Ferris, she's Wilson's secretary, gave me the money. I brought it home for safe keeping.' She scoured her hands dry on her pinny. 'More tea, sergeant?'

Barlow nodded. She poured and the talk drifted onto other subjects; mainly her children, now married with their own families. Eventually she stopped herself talking and sat silent, wishing him to go.

'Where was it then?' asked Barlow somewhat rudely.

She touched the table. 'It was wrapped you know.' She rushed to a drawer and pulled out a handful of folded brown bags and lengths of string. 'I counted it, then Wilson checked it for me. I put it in a bag and he tied it with string....' Her voice faded away, she looked confused.

Barlow pulled out his notebook and added the name of Miss Ferris to his list. He said, 'So you counted it, your husband checked it and you're still not sure if there was six hundred pounds in that bag or not.'

'No,' she said. She was very firm about that.

District Inspector Harvey was as imperious as a Roman bust behind his pedestal desk. 'Where have you been?'

'A remand hearing, sir.'

Harvey's hand came out. 'I want to see your notebook.'

Barlow took his time about handing it over. Harvey flicked through the pages, stopping occasionally to puzzle at Barlow's crude pencilling. Eventually he said, 'Your interview with Mrs Kyle, it's not here.' He pretended to shuffle through the papers on his desk. 'And your report, where is it?'

Barlow kept his eye on the portrait of the young Queen on the wall behind Harvey; he thought the lass was letting things get out of hand with people like Harvey in control. 'The lady wasn't willing to make a complaint.'

'Well she is now.'

Harvey was like a fighting cock with ruffled feathers. Barlow said, 'You've got a quare tongue in your head, sir, I got nowhere with the woman.' He was wondering why Mrs Kyle had changed her mind.

Harvey subsided slightly. 'Young Anderson is in custody. Take a statement then charge him.'

Barlow nodded and left. He tramped into the front office. 'Where's Robby Anderson?'

'In the cells, Sergeant,' said Constable Jackson.

'Well get him out of there,' roared Barlow, and stomped on to the interview rooms.

There were voices echoing out of one of them. He looked in, and saw Mrs Kyle and another, younger, woman sitting with a china cup of tea in their hands. Mrs Kyle coloured and looked away. The other woman needed glasses, Barlow noted, because the look she gave him was part squint.

'Excuse me, missis.'

'It's Ferris, *Miss* Ferris.' said the younger woman.

'The secretary?' asked Barlow.

Miss Ferris's head went higher and the nose sharper. 'If it's about the money? I can assure you that it was all there. I counted it out to Mrs Kyle myself.'

'She did,' said Mrs Kyle, taking courage from Miss Ferris's confidence.

Barlow marked Wilson Kyle's lorries down for special attention the next time he saw them obstructing traffic then went on into an empty room to await Robby Anderson.

Robby slouched in, red eyed and angry. He glared at Barlow. 'I want my solicitor present.'

'You don't have one,' snapped Barlow, annoyed at the boy's surliness. He was nothing like his mother. She had been quite a girl in her day, but had suddenly married Robert Senior and settled down. Barlow told Constable Jackson to get tea and the duty solicitor. With the help of the solicitor Robby made a full statement, right down to Moncrief and the train set.

Harvey was not amused, he and Moncrief played golf together. 'If this is read out in court people will think….' He struggled for the right words and ended up wagging his finger at Barlow. 'Your hand is in this somewhere, one slip and I'll have your pension.'

Barlow met Edward by arrangement in the Bridge Bar; even for a Monday the place was empty. Early as he was, Edward was there before him, sitting huddled on a stool.

'Bugger,' sighed Barlow. Once Edward started drinking there was no getting sense out of him.

The barman was in the process of bringing Edward a refill, a pint glass of what looked like well-watered green paint. 'What's that?' demanded Barlow.

'Water and lime cordial,' said Edward, and looked sick. 'I have information to impart that may have a profound significance in relation to your ongoing investigation.'

'What have you found out?' asked Barlow, relieved that Edward had remembered the meeting, let alone what it was about.

Edward pushed his drink away with a shudder. 'Fifty-five pounds goes missing while in the custody of young Robert, and Moncrief's son pays fifty-five pounds for the train set.'

'And?'

'The deposit on a new house in the Model Estate is fifty-five pounds.'

'Begod, you're right.' said Barlow. He opened his notebook at the list of names and put it on the counter. Edward topped it with a site plan of the Model Estate, a name was printed against the outline each house.

Barlow lined up the papers on the counter before him; he squared the edges off neatly. 'We could do this quick, but I think there's more than one story here.'

Edward nodded and licked dry lips.

Barlow looked at the map of the Model Estate and blinked in surprise. A Miss Ferris had bought one of the houses. He sensed that the lives of the people on his list were all there, interconnecting, if only he could work them out. After a time, he caught the barman's eye and ordered two pints. When they came he held them well out of Edward's reach. 'You're a man who's always walking about and seeing things. That gives us a third list, doesn't it?'

'You're a smart man, Mr Moncrief, that's why I came to you,' said Barlow.

He was standing in front of Moncrief's desk; Moncrief sat behind it. The desk was paper cluttered and the room was dull, in keeping with the solemnity of the work.

Moncrief was tight and dressed sharp, and thin; he was as mean with food as he was with anything else. 'I have already told you, sergeant, I cannot possibly discuss a client's affairs with you.'

'I'll tell you, sir, you don't even have to nod.'

Moncrief said nothing. Barlow took that as agreement. 'They way I see it is this. Wilson Kyle has been having it off with his secretary, Miss Ferris, and she wants something more than promises - a house, say.' Moncrief's head came up at that, Barlow pretended not to notice. 'So she puts a bit of pressure on Mr Kyle, something between blackmail and enticement, to come up with the deposit. He daren't write a cheque in case his wife notices - she suspects you know, wives are quick that way - so he slips the cash out of the lodgement and parcels it up again.'

Moncrief tapped his fingertips together. 'Supposition, sergeant, pure supposition on your part and I must warn you…'

'Not quite, sir.' Barlow pulled out his notebook and sought a well-pencilled page. 'Miss Ferris paid the builder the deposit in cash and he kept some of it, wrapping paper and all, a bit of a tax dodge, if you see what I mean. Now if we were to find Mr and Mrs Kyle's fingerprints on the wrapping paper or on the notes themselves, we'd have a bit of prima face evidence. Then there's the question of the train set.'

Moncrief was very still; his fingers had stopped tapping. 'What about the train set?'

Barlow was at his blandest. 'Oh, just how things tie together, sir. Who knew what.'

'Like?'

'Like how Miss Ferris didn't pay the builder personally, her solicitor did. You.'

Moncrief jerked to his feet. 'One word, sergeant, one word out of place and it will cost you your pension.'

Barlow shook his head. 'Everybody wants my pension, it's not that great.' He straightened his jacket and prepared to leave. 'Another two hundred for the train set and it needn't be mentioned. Theft, you see, a forced sale of goods.'

Moncrief's finger never wavered. 'Out!'

Barlow put his cap on, from here on the interview was official and to be reported. 'Mrs Kyle keeps too tight a hand on the money for Mr Kyle to slip out the price of a house, look at the trouble he had coming up with the deposit. But a solicitor always has money, and you and Miss Ferris have been seen together quite a lot. Around Curles Bridge, sir, where the lovers go.'

Moncrief sat down again. There was a sheen of sweat on his forehead.

Barlow said, 'Clear the boy's name, and give him the money to see him through university. Help him to become something better than a solicitor's clerk.'

Moncrief sat tight-lipped while a dray horse and cart rattled past the window. When the sound faded he said, 'I admit to nothing. But if… my client… did as you asked, that would be the end of the matter. I have your word?'

Barlow assessed Moncrief. He didn't have Robby's eyes. Neither did Edward for that matter nor his mother, he was sure of that now. His mother, who, as a young woman, always had time for a constable on night patrol.

Barlow said, 'Robby no longer trusts people. How do you make good that?'

PROOF

If the Dunlops were out to annoy they were doing rightly because they were in the pub used by Sergeant Barlow, the Bridge Bar. They followed Barlow in and sat: loudmouthed, shouting their orders, and making a point of not noticing him. 'And none of that bloody Kirktown piss-water,' roared Geordie, the elder of the two. He threw a scattering of pennies, threepennies and sixpennies onto the table.

For duty's sake Barlow was sampling the Guinness while he talked to the barman about licensing hours and rowdy customers. He took his pint over to the Dunlops' table, sat down and loosened the collar of his uniform jacket to let his neck breathe. 'So you're going to jail?' he said. 'Nicking cattle?' he said.

The Dunlops had got away with a lorry load of whiskey at Christmas, but Barlow was happy with the lesser charge of stealing cattle. It wasn't his whiskey and, anyway, it was almost traditional, the Dunlops taking enough from the brewery to see them over the festive season. But the last Christmas, new broom District Inspector Harvey, had interfered and caught some of the younger Dunlops in the act. The lorry load of whiskey had disappeared the same night.

'Us in jail? You'll be lucky,' said, Geordie.

'And we'll have you for police brutality,' said Geordie's son, favouring the elbow Barlow had truncheoned during the arrest.

Barlow shook his head at the younger Dunlop. 'You got it bending the elbow with stolen whiskey.'

'You've no proof,' said Geordie.

'And you've had more than enough,' said Barlow. He looked at the bloated face of Geordie and tried to find something of the man who had served with him in the Kirktown Rifles during the war. The 'Priests' were the local regiment and the town was thick with men who walked at double time; some of the police too, though that spoiled them for strolling the beat.

Barlow said, 'See you in court.' He buttoned up his collar and left, and failed to see Mr Edward Adair, gentleman, and tramp by profession, hanging round the door, even though Edward coughed and made a play of not having enough coins for a pint.

Constable Jackson fell into step beside Barlow. 'Those Dunlops are a bad lot, Sergeant.'

Barlow shook his head. 'They just charge, somebody has to goad them first.' He pointed to a snarl-up in the traffic; a Martin's lorry was double-parked at the Co-Op while unloading. 'Get that,' he said, and stood thinking to himself. The Dunlops were up for the attempted theft of three cattle belonging to the Chairman of the Police Authority. A dozen cattle had disappeared before that, all of them prime bullocks, making it appear that the Dunlops were stealing to order.

He wondered who could dispose of two or three head of cattle a week: not the local butchers, most of them wouldn't do three beasts in a month. And, for the rest, the dealers would soon notice if orders were down by that much. He thought about it on his way back to the station. At least that's where he ended up having missed three pubs on the way.

He rounded on Jackson. 'Three pubs! You're supposed to tell me things like that.'

'But you…'

'Were waiting for you to react. You'll have to work on it, son.'

Jackson went off in a huff. Barlow dandered into the kitchen and heated up the tea in the pot. He was dozing over an article in the paper about the 'Priests' being in action again when Jackson came in. Barlow let him clear his throat and shuffle his feet a few times before he growled awake. 'Son, son, son, when will you ever learn? You ignore inspectors, you keep clear of district inspectors, and you never disturb a sergeant when he's thinking.'

'Sergeant, Mr Harvey would like to see you.'

'And you said I was here?'

'Well…'

District Inspector Harvey was at his desk with a glass of whiskey and a loose cheque in front of him. The cheque for seven pounds was payable to Solicitor Moncrief; Barlow made a point of noticing.

Moncrief sat comfortably in a leather-covered armchair. His nose was pinched, it could have been the taste of the special reserve in his hand, but it was more likely Barlow's presence. Barlow stood. There never was an invitation for him to sit in Harvey's office.

'The Dunlops!' began Harvey.

Moncrief gave a dry laugh. 'They wouldn't even make sergeant, no drive you see, no leadership.'

'You, could be saying that, sir,' said Barlow, to encourage him on, but Harvey glared at Moncrief for interrupting and he remained silent.

Harvey sat more upright in his chair; he checked his collar and tie. 'The Dunlops, through Mr Moncrief here, are willing to give us a name in exchange for a sympathetic review of the charges pending against them.'

Barlow scratched along his jaw line. 'The Chairman won't like it.'

Moncrief gave a snort of derision. 'He'd hang them if he could.'

Barlow looked him full in the eye. 'There's worse than the Dunlops, sir; at least they're honest thieves.'

Moncrief went still, his eyes burned.

Harvey said, 'Quite right, Barlow. The man who bought the cattle and the drink off the Dunlops, that's the real rogue in this case.'

Moncrief stood up and lifted the cheque from the desk. He paused in the doorway. 'Think about it Mr Harvey, my clients are only pawns in a bigger game.'

For some reason Barlow got the idea that Moncrief's smirk was directed at him. He stared at the young Queen's portrait on the wall behind Harvey; the brightness of the day against the glass made it look faded. He thought a drink wouldn't do either of them any harm.

'Did you have a good game, sir?' he asked, as the door closed behind the solicitor.

Harvey snapped. 'That man! He suggested that we play for the usual pound; I thought he meant for the round. After the game he said I owed him seven pounds, one for each hole down. I never play for that sort of money, not even at Royal Portrush.'

'Bad luck, sir.'

Harvey's lips made a tight thin line. 'There'll be no deal with Mr Moncrief. Now we know there's a Mr Big I want you to find him.

Five years he'll get, five years.' He almost drooled as he counted off the charges. 'Receiving stolen goods, conspiracy....'

'I'll see to it, sir,' said Barlow, and went back to his paper and the pot of tea.

The next morning Barlow detailed Constable Jackson to patrol Castle Street and note the names of everyone going into Solicitor Moncrief's office. At mid-morning, Barlow did his rounds and found Jackson leaning against a gable wall. Jackson quivered to attention when Barlow barked into his ear. 'You'd pass for a corner-boy any day.'

'Yes, Sergeant. Sorry, Sergeant.'

Barlow looked down the street. It was empty, not even a parked car in sight. 'Well?'

Constable Jackson had only two names for Barlow. 'The Dunlops, they're still with Solicitor Moncrief. And they went in quiet, which isn't like them.'

Barlow scratched at the stubble on his chin. 'How did they arrive?'

'Walking,' said Jackson, surprised.

'Walking? And them had a lorry for stealing cattle with.' Barlow scratched again, it was the nearest he could get to the itch in his brain. 'When they come out, see where they go.'

He took a turn round the town then went back to the station, going in by the yard, and stood for a time looking at the Dunlop's impounded lorry. After a while he muttered to himself, 'I wonder who owned it before? Maybe still does?'

He went indoors, just in time to catch Constable Jackson sneaking into the squad room. 'Well?' he demanded.

'Nothing, Sergeant. They went over Curles Bridge so I came back.'

'Did they talk to anyone? Have a drink?'

'No'

'Did you wait?'

'Half an hour, not a sign of them.'

Barlow grunted and went to make his telephone enquiry about the lorry. Harvey was out, hob-nobbing at the town hall, so he made it from his room. While he was there, sitting at Harvey's desk, he checked the drawers for anything of interest and found nothing. 'Like the man himself,' he muttered, and went off for his lunch.

When he got back Constable Jackson was hanging over the counter anxious to grab his attention. 'Sarge… Sergeant, the tramp, Mr Edward, he's the Mister Big behind the stolen cattle.'

'Balls!' roared Barlow.

Jackson took cover behind the Incident Book. 'He admits it. And Mr Harvey is all pleased with himself for solving the case.'

'You couldn't leave the place a minute,' muttered Barlow. He took the Incident Book off Jackson and gently replaced it on the counter. 'So tell me.'

Jackson relaxed visibly. 'The Vehicle Licensing Department returned your call. They say the lorry in the yard is registered in the name of Mr Edward and gave his address as Curles Bridge.'

'And you let Inspector Harvey see the message?'

'I was on lunch, Sergeant,' said Jackson with studied innocence, and added. 'Mr Edward admitted to everything right away, and he said the drink stolen at Christmas was him as well.'

Barlow squared off the Incident Book against the counter edge. He looked at Jackson, with a tilt to his head. 'And he believed him?'

'Well… yes, Sergeant.'

'He would,' said Barlow.

Edward was in the cells, the one held reserved for him during periods of what he gracefully referred to as inclement weather. Against all regulations the door was on the push.

'It's very good of you to call,' said Edward.

'You're a bollocks,' said Barlow.

'My dear chap.'

Barlow leaned against the wall and folded his arms. 'Where did you get the money to buy a lorry?'

'The family trust, as I've already told Mr Harvey.'

Barlow grunted. 'Is that the same trust that funded your tar and paper shack under Curles Bridge?'

Colour popped into Edward's sallow cheeks. 'There's no need to be offensive.' He sat stiff with pride. 'Some of our wartime accommodation was of a lesser standard.'

'Where you're going they lock the doors. How will you like that?'

Edward swallowed and had no answer. They looked at each other, – Edward from his seat on the bunk, Barlow still leaning against the wall. Edward ran a hand over dry lips.

'There's no booze there, either,' said Barlow. He straightened and brushed dust off his sleeve. 'So what's this all about?'

Edward coughed gracefully into a knuckle. 'The Kirktown Rifles…'

'What have the 'Priests' got to do with it?'

'They're in action, my dear chap. Borneo, one of those funny little places at the end of the Empire.'

'And?'

'The Comforts Fund was running done, and when you're in the jungle those little extras make all the difference.'

Barlow finally believed him. All the cattle had been stolen from rich farmers or businessmen, like the Chairman of the Police Authority, who farmed as a hobby. He said, 'I hope you enjoy the next five years just as much.'

Suddenly Edward looked strained and twitchy. 'I thought perhaps six months…'

'You and the Dunlops,' said Barlow, brutally.

'The guilt is mine, I compelled them to cooperate.'

'Two of the tightest men in the country?'

Edward drew himself up to his fullest height. 'Moral pressure, my dear chap. Blood counted in the end.'

There was the sound of feet in the corridor: one set heavy and authoritive, the other tiptoe and anxious. The Chairman of the Police Authority appeared in the doorway. 'Bloody hell, Edward,' he said.

Edward stood up and extended his hand. 'Charles, how good to see you.'

The Chairman refused to shake hands. 'You're a bollocks!'

'Sergeant Barlow has already expressed that compliment.'

The Chairman turned on Barlow. 'You're supposed to keep him out of trouble.'

Barlow shook his head. 'He'll tell another story when he sobers up.'

'He's not drunk,' said District Inspector Harvey, indignantly, from behind the Chairman's elbow.

'God forbid that I should,' said Edward.

'Sober that is,' said Barlow, and explained to the Chairman about the 'Priests' and the Comforts Fund.

The Chairman went red with anger. 'But I contributed.'

'Three bullocks?' asked Edward.

A row started between Edward and the Chairman. Or rather Edward stood silent while the Chairman shouted his points: he and Edward had been friends from primary school. He, the Chairman, was a former soldier, and it was all very embarrassing having to prosecute his old Company Commander. 'And what do you mean the Dunlops didn't take my whiskey? The country knows it was them.'

In the end he turned on Harvey. 'Send him home.'

Harvey vibrated with indignation. 'He has been charged…'

'I withdraw the complaint,' roared the Chairman. He glared at Edward. 'This means that the Dunlops get off as well. I'll never forgive you for that.'

Harvey was white to the gills. He flashed a look at Barlow, daring him to smirk. Barlow kept his face straight and, anyway, his brain still itched. He said, trying to mend a few bridges, 'Mr Harvey, you said something about the drink.'

Harvey hadn't, but he tried to remember anyway. 'Remind me.'

Barlow spoke slowly because he was still working on that itch. 'About the cattle being unimportant, but the drink being serious.'

'You're bloody right,' said the Chairman. He brought his hands together like he was wringing a Dunlop neck. 'Half a year's profits gone like that.'

Barlow was still thinking out loud. 'Which rules out Edward, he can't bring himself to hurt people. Isn't that why he drinks?'

'Certainly not!' said Edward.

Barlow hardly heard the indignant reply. He walked out of the cell, his face set in a pout of thought.

The Chairman shouted after him. 'Tell me who bought the whiskey and I'll destroy them.'

'Aye,' said Barlow, to himself. 'Like the cattle and the butchers it never showed.' He stopped in mid-stride, almost in pain from the cessation of itch. He now knew where the cattle had gone. Martin's

Produce had the contract for supplying the army. Some day, in his own good time, he would sort out Sam Martin.

When he got to the front desk the Chairman was still shouting at Edward and Jackson was agog with curiosity.

'It's all your fault,' Barlow told him as he lifted the keys for the lorry off the hook. 'Come on,' he said, and walked off at double-time.

Jackson trotted after him. 'Where too, Sergeant?'

Barlow stopped so suddenly that Jackson nearly cannoned into him. 'Like you said, Curles Bridge.'

'I did?' Jackson looked confused, but followed him out into the yard and got into the lorry when told to.

The lorry hadn't been started in weeks, its engine churned slowly and Barlow dared it to let him down. It coughed into life and off they went through the town and across Curles Bridge, over the river and the dry arch where Edward lived in his tar and paper house. Beyond that was a copse of trees, the remains of the old Adair Estate. Barlow took a muddy path through the copse.

The lorry lurched from pothole to pothole. Jackson shouted over the noise of the engine and the labouring springs. 'Where are we going, sergeant?'

Barlow said, 'If the Dunlops came here there was a drink at the end of it.'

He followed another, muddier path, and pulled up at a row of abandoned two up, two down houses; one doorstep was marked with fresh mud.

The front door was stuck. Barlow kicked it open and walked in.

Geordie Dunlop and his son sat on up-ended whiskey cases, sharing a bottle between them, and behind them were row upon row of whiskey cases. Barlow made himself comfortable on a stray case and looked from father to son. Jackson filled the doorway as best he could and drew his truncheon.

'Put that away,' said Barlow to Jackson, then he loosened the top button of his uniform jacket and massaged his neck. Geordie found a glass at his feet. He swilled it out, poured a generous measure of whiskey and handed it to Barlow.

Barlow passed it on to Jackson, saying, 'Drink it, son. Don't insult the man.'

Jackson struggled to get the neat spirits down. He shook his head and blinked and stumbled without moving his feet. Barlow accepted the refilled glass and sipped at it. He said, 'I'm off duty until I finish this.' He took another sip and leaned against the stacked cases. 'I was thinking, if you can't sell the stuff because all the barmen are too frightened of the Chairman to buy it, and if you can't drink it without killing yourselves then maybe you should hand it back?'

Geordie raised the bottle to his lips. Barlow leaned forward and pulled it down again. 'Six months in jail would dry you out.'

The son said, 'We'd get five years, that's too much for da.'

'Oh I don't know,' said Barlow. 'The Chairman will be pleased to get to get the whiskey back. Maybe we could get a deal going. Drop the charges for stealing the cattle, and say you stole the whiskey when you were drunk and didn't know what to do with the stuff when you sobered.'

'You'd do that for us?' asked Geordie.

'I would.'

'Why?' asked the son.

Barlow finished his glass of whiskey. 'Solicitor Moncrief is trying to get at me through you.' His voice dropped. 'And don't we all owe Mr Edward for getting us through the war?'

'No cattle?' asked Geordie.

'Charges withdrawn as if it never happened.'

Geordie lumbered to his feet and lobbed the first case of whiskey to Jackson. Jackson and the case somersaulted out the door.

It was a month later and the Dunlops had just been put away for six months. 'A foolish drunken escapade,' the magistrate called it, and was mindful of the voluntary return of the whiskey when setting the sentence.

Jackson was waiting for Barlow when he came back from his lunch. Jackson looked nervous as he lifted a Heinz beans case onto the counter. 'Mrs Dunlop left this for you. She told me to say thanks from Geordie.'

Something inside the case clinked. Barlow looked in and found three bottles of Kirktown whiskey.

Jackson said, 'Sarge, that's…'

Barlow pushed a bottle into Jackson's hands and kept the other two for himself. He shook his head. 'Son, you've no proof.'

HANG-UP

It was raining in Kirktown, straight down rain like lines drawn on a page. A white Jaguar car splashed Sergeant Barlow and his bike on the way past.

Barlow wobbled to a halt. He sat with one boot in the puddle and water streaming off his uniform while he shook his fist at the car. 'God curse you, Sam Martin, but you're due for a land.'

And he was too. Barlow had enough *picks* against Sam Martin and his provision lorries to make a decent sized scab. He pushed on because a busybody had reported young boys for playing football near a millstream, and it foaming over the edges. He found the boys and confiscated the ball.

The owner of the ball, a stocky wee fellow in a red jumper hooded his eyebrows and set his mouth in a twist of temper. Barlow tried to clip his ear for him, but the bike and the ball got in the way. 'Go to the People's Park,' he told them. 'Drown in the pond there and save the Council a fortune in duck food.'

The boys headed off, the red jumper last of all, and Barlow was left with a pleasurable freewheel back into the town centre. Except for the ball that is: a nuisance to carry and greasy with mud. He punted it away and got the red jumper right between the shoulder blades. He rode on with his mind made up. Lord Mayor or not, he was going to do Sam Martin.

His chance came sooner than he thought. Sam Martin had parked his white Jaguar outside the Town Hall, on the corner of High Street and Castle Street. A ticket would have been washed away so Barlow thoughtfully left it for Martin at Reception. Another short freewheel took him back to the station, where he found young Constable Jackson manning the front desk and hopping from one foot to the other in his anxiety. 'Sarge, the mayor's been on the phone. He shouted a lot.'

'Did he now?' said Barlow, and went to dry out in the kitchen only to find Mr Edward Adair, gentleman, and tramp by profession, making himself comfortable in Barlow's own chair. Edward's old army greatcoat steamed gently in the heat from the range.

'What are you doing here?' asked Barlow.

Edward smiled. 'My dear chap.' He sat on, and made himself more comfortable against the backrest.

Barlow jerked his thumb. 'Shift.' Edward stood up and looked on indulgently while Barlow undid his laces and wriggled his toes.

Edward said, 'You're grouchy, water always has that effect on you.'

'Whose fault is that?' asked Barlow. He looked up. 'What are you doing here any way?'

'My accommodation is… unapproachable.'

'Good.'

Edward sniffed. 'One should not find humour in the misfortune of others.'

'I told you not to put it under the dry arch. Didn't I?'

'You did, my dear chap, you did.'

'Well you can't stay here,' said Barlow, and shouted at Constable Jackson to get them both something from the café across the way. He went off to change before somebody told District Inspector Harvey that he was back in the station.

He ran into the inspector right outside the door. Harvey's normal walk was a creep. 'The mayor has been on the phone, he's very annoyed,' began Harvey.

'Just doing my duty, sir.'

'But he's the mayor?'

'I can't help that, sir,' said Barlow.

Harvey recovered quickly. 'Well anyway, one arch of the bridge is jammed; we're on flood alert.'

Edward appeared at the doorway. 'But my house?'

'Shack,' said Barlow.

Harvey dry-coughed to regain their attention. 'The river is pouring through the lower end of the town.'

Barlow said, 'There's a lot of wee homes down there.'

'And businesses.' Harvey vibrated with indignation. 'The workers won't come in. They want to save their own things first.' He waited for Barlow to say something. He didn't so Harvey added. 'On you go, I'll collect the rest of the men and follow.' He flapped a hand in the direction of Edward. 'And take that *person* with you.' He crept on at a rush.

Barlow sighed. 'And the wife's got new lino in the kitchen, there was no talking to the woman.' He pulled on his boots and yelled for Jackson. 'Come on, son, we need your finger for the dyke.'

A section of the riverbank had given way and the army was on the job with sandbags. Dirty river water slopped down the Derry Road, curled round the shading trees and forced its way into the town cinema; they said the rats were standing on the seats. At the top of the road, at the roundabout, the river wheeled right and disappeared in among the mill houses. Those who had nothing to lose stood with their toecaps at the water's edge and talked among themselves.

Barlow looked at the scene in disgust. 'Money doesn't pay you at times,' he said, as a crowd of hapless women spotted his uniform and converged on him; their cries of despair coming flat off the heavy water.

Constable Jackson said, nervously. 'Sarge, shouldn't we do something?' He put Barlow between himself and the women.

Barlow grunted. 'Do what? The river for obstruction?'

The water rose and lapped over his boots as he spoke. He strode purposefully through it and only stopped when the women surrounded him, swamping all but his head and shoulders.

'What's bothering you now?' he asked.

They all spoke at once. Their houses were underwater, some to the letterboxes of the dark stained doors, and one woman claimed that the baby she carried had a cough and would catch her death.

Barlow tickled the baby under the chin, it gurgled in delight.

'Is anybody drowned?' he demanded.

There wasn't, and there was an abbreviated crossing of bosoms and 'God forbids.'

'There you are then,' he said, with satisfaction, and gave the baby a final tickle. But his face dropped when Edward surmised that the Bridge Bar, Barlow's regular drinking hole, was bound to be underwater. 'It's a bad do,' moaned Barlow. His eyes flashed over the women. 'And you lot cackling as well doesn't help.'

He told Jackson to round the strays up and take them to the

Kingdom Hall. 'They're bound to have an upper room.' Then he tramped upstream to his own house. The linoleum was still above water, though the wife was worrying about it, and her matches were too damp for lighting the gas. She harped back to the lino. 'And you bought it for my birthday, a special treat. I told you the old one would do another year.'

Barlow interrupted her. 'Would you quit worrying, woman.'

He took a few minutes our of saving the town to see her and the canary safe upstairs; he reckoned the time well spent because a bit of petting and reassurance always ended up as something special on the tea table. He gave her a box of matches out of his own pocket and went on.

Young Constable Jackson rejoined them, and he and Edward splashed alongside Barlow. They both looked cheerful, Barlow wasn't because his boots were full of water. Given time his feet would warm, but just then he felt miserably cold and in no mood for the mayor, Sam Martin, to come nosing down Prince's Street in his white Jaguar.

'Sir,' he touched his cap with his finger.

'Well?' demanded Martin, out the driver's window.

'Thank you, sir, yes,' said Barlow.

Martin flushed with annoyance. 'I mean here. Why aren't you doing something? You're handy enough when it comes to parking tickets.'

'I'm the advance party, sir, doing a recce. There'll be more people here in a minute than you could shake a stick at.' Barlow pretended to look around. 'Where are the council men?'

Overtime and budgets came out in a splutter of words. Those men the council could afford to pay were laying sandbags in Queen Street.

Barlow said, 'There's nothing there but your produce store.'

Martin's face went rigid. 'I'm not answerable to you and, anyway, it's dangerous around flood water so that makes it a job for the police.'

Barlow nodded, and he, and Edward, and Constable Jackson joined the line of people carrying things from threatened houses, and up the hill. Sam Martin stayed dry and used a chamois to rub at the trace of Barlow's hand on the Jaguar's paintwork.

District Inspector Harvey came bowling up with a carload of constables. Barlow gave them their orders then suggested that Harvey should give the mayor a lift to the town hall from where they could organise relief. Martin liked the idea of being whisked away in a police car with the bell clanging, and swelled with importance as he got in beside Harvey.

Barlow watched them go. He muttered. 'You'd make great staff officers. You talk a good war.'

Sam Martin had left his own car locked and obstructing people trying to pass with their bits of furniture. Barlow noticed that a window of the car was partway down. He got Jackson to slip his thin arm in and pull the catch, then he freewheeled it downhill and out of the way.

The rain had stopped at long last and somebody handed him a cup of tea. There was nothing in it to give him fortitude, but he drank it anyway and went off to the next trouble spot. They came to a point in the road from where they could see the ever-expanding river. The bend in the river was now an expanse of water, with roofs of houses doubling as channel markers. From the streets, a leaden stream of water, dotted with furniture and a galvanised bath, cascaded down the riverbank and boiled through the remaining arch. No bodies Barlow noted with relief; they would take some fishing out in that kind of weather.

Edward's house was still there, under the normally dry arch, but well waterlogged.

'The devil takes care of its own,' said Barlow, and, at the behest of a concerned lady, who dripped jewellery and haughtiness, told Jackson to rescue a cat and carry it to safety. 'Don't drown yourself in the process,' he roared, as Jackson charged into the water without giving thought to depth or current.

Edward said, 'I remember a young army sergeant of mine…'

'I was never that young,' denied Barlow. He took the cat off Jackson and presented it the lady. The fact that she was Mrs Martin, the mayor's wife, and nowhere near her home, was something Barlow put at the back of his mind to look into another day. There had been rumours… He accepted the five bob tip with studied nonchalance and thought about going back to the station to change his socks.

A workman, soaked to the waist and dripping sweat, came running up. A tangle of trees had been spotted in the river upstream. If they blocked the remaining arch the rest of the town would disappear underwater.

Edward said, 'The bridge could give way.'

Barlow visualised his drinking hole, The Bridge Bar, slowly disappearing under the backed-up water. 'Too late,' he said.

One minute the street along the river ran empty, the next people thronged, anxious to see an even bigger disaster. The more curious of the gawkers followed Barlow and Edward out onto the bridge to get a better view. The water was like a molten hemp rope, turning and twisting into itself, the strands separating as they came to the bridge then replaiting themselves beyond the arches. The bridge was metal, bolted onto the pillars of an older stone-arch bridge. It sang under the force of water, whose speed was that of Edward late for opening time.

The trees bore down on the watchers at a more sedate pace, heading for the already blocked arch. At the last minute one broke away and was carried along by the current towards the remaining arch.

Edward stood with his head tilted up, the way a short-sighted person might to improve his vision. In his case it was pride, breeding responding to the family motto: *First into danger*. 'Suggestions?' he requested.

'That you bugger off,' said Barlow.

'It's dangerous. You shouldn't be here, sir,' said Jackson.

Edward patted his arm. 'Ah, the faithful retainer.'

Barlow glared at him, and while he glared the trees struck the bridge. It shuddered under the impact; there was a sharp metallic crack followed by something whistling through the air. The three men shuffled their feet to keep their balance; everyone else had run to the safety of the riverbank.

'Well, well,' said Barlow, looking at what could have been a wall of scrub growing above the parapet. Downstream there was a definite reduction in the flow of water so the tree was plugging the arch. He went from parapet to parapet, measuring the debris piling up against the bridge, and even as he looked the water gulped more of the tree under the bridge and a tangle of roots appeared. Debris piled higher and faster, and downstream the flow of water fell. In half an hour, an hour at most, the two arches of the bridge would be completely blocked and the town lost. The bridge no longer sang, but groaned and, again, something cracked: sharp and loud.

The third arch, the normally dry arch, under which Edward lived, did its best to cope and disposed of its own debris to keep itself clear.

Edward watched his tar and paper hut swirl away and gradually sink into the water. 'Easily replaced,' he said.

Barlow grunted. 'I'm the one who has to do the hammering and nailing, you're totally handless.'

'My dear chap.' Edward leaned so far over the parapet that Jackson had to hold onto his belt to stop him from falling into the water. Eventually he pushed himself back. 'An axe,' he told Jackson.

'And a rope,' added Barlow, and told Jackson to run.

The tree's spreading branches held it in place. If they could be chopped away, it would clear the bridge and save the town from further flooding. Barlow shook his head at Edward because it meant standing on the tree while swinging the axe, and depending on the rope to pull them clear when it started to move.

'I'm open to alternative suggestions,' said Edward. A wind had got up and his greatcoat flapped to and fro. He stood, oblivious both to it and to the fine spray it carried off the river.

Barlow looked towards the bank. The lake of water in the town had grown; it had to be near his house and that new linoleum.

Jackson came back with a fire engine packed with men. The Leading Fireman had a look for himself and backed off. 'You've got to be joking.'

'Certainly not,' said Edward. He dragged the man back to the edge and pointed down. 'Look, we stand on the tree trunk and use the axe to weaken the branches. As soon as it begins to move you haul us up.'

The man spluttered and argued about not wanting their deaths on his conscience. He continued to argue while Barlow tied a rope around his own waist and Edward's, and swung himself onto the parapet. Edward joined him and they gave it a moment while the firemen swung their extending-ladder out over their heads and secured the ends of the rope to that.

Barlow spat into the molten stream below. 'I hate bloody water.'

Edward laughed. 'You stopped taking it in your whiskey after that last time.'

'Do you blame me?' asked Barlow, as Edward swung himself down onto the trunk.

The Leading Fireman tried to stop Barlow, who shook his head. 'Mr Edward got me through the war, it's my job to get him through the peace.'

He jumped down, the tree trunk flexed under his weight and he found himself in water again. This time the force of it threatened to whip his feet from under him.

Barlow pulled on the rope and shouted for a bit of slack. He waited for the turntable to drop the ladder a few feet before hefting his axe, but deferred to Edward for the first swing. Wasted as he was with drink, Edward had a wiry strength and both axes bit deep. The flood water rose quickly from their ankles to their knees and they pressed back to back for comfort and support.

Barlow stopped and drew breath. He had to shout to make himself heard over the noise of the water, 'Maybe us being up to our oxters in water in that cellar, and the bomb ticking, wasn't so bad after all.' Edward was laughing openly as he worked. Barlow shook his head. 'Even as an officer you were a mad bugger.'

He took two more swings and suddenly the tree was gone, sucked through the archway, and he was underwater, being dragged into the darkness under the bridge. He closed his eyes, and thought of the wife and the canary safe upstairs. Debris went past him, touching but doing no damage, then he felt himself being pulled up, scraped over the parapet and dumped onto the pavement. He landed in a heap. Edward was there before him.

'You got here first,' said Barlow. It could have been an accusation.

Edward brushed a fleck of dust off the shoulder of his greatcoat. 'My dear chap, rank has its privileges, even in this egalitarian age.'

'You mean, whiskey floats.'

The Leading Fireman leaned over Edward and saluted. 'Are you all right, major?'

'As usual, I'm alive,' said Edward, then he threw his head back and bellowed with laughter. Barlow laughed with him, as did Jackson and the firemen, and the river roared through the archway. Its level dropped.

They got to their feet and walked through the town, heading for the police station and the warmth of its kitchen. By keeping to the pavements and detouring a bit they were more splashing through the floodwater than wading.

The Bridge Bar was open. At least the barman was in his fishing waders making a show of sweeping the water out. They went in and lined up along the bar.

'Whiskey,' said Barlow.

He looked out the side window and up the stretch of road towards Princes Street. Sam Martin's Jaguar was hard up against a telegraph pole. The driver's door hung twisted off one hinge and mud smeared the bodywork. The windscreen was broken.

Edward was still tense from the danger they had faced. It was Major, The Honourable Edward Adair, formally of the Royal Engineers, specialist in unexploded bombs, who snapped. 'Did you put the handbrake on and the engine back in gear?'

'I did,' said Barlow, to the question about the handbrake, but not how much. He watched as his toecaps appeared out of the receding waters, and thought things were looking up. The linoleum and the wife should be all right.

The barman put three glasses on the counter and poured the whiskey. He lifted a jug. 'Water?' he asked.

Barlow shuddered.

PLAYING WITH FIRE

CICADA

'Franklin!'

Gordon snapped awake.

Franklin was standing outside the auberge looking up at the mountains. Honey, his wife, was panicking in case non-existent traffic knocked him down.

Franklin said, 'The next bit's gonna be some drive.'

'Maybe we shouldn't. The man at the hotel said people have died.'

Franklin laughed. 'Drunk kids, a couple of skinny-dippers.' The laugh was wheezy; he used a steroid inhaler to clear his bronchial tubes. Then he looked in under the arboretum that laced the tabletops with shade. 'Well I'll be. Gordon.' He rushed over.

Gordon sat on, feeling sick in the stomach. He said, 'I left you in charge.'

Honey's smile never reached her eyes. 'A good man finds time for holidays, the others are sent.'

Franklin said, 'I keep in touch by phone, no problem.'

The anger spilled out of Gordon. 'All the problems you solved, did you create them in the first place?' Somewhere above his head cicada sang, the noise was like a corncrake clearing its throat.

'Bonjour Monsieur, et Madame?' It was the owner, Madame Defarge, with menus.

'Nothing,' said Honey. 'We only stopped to use the powder room.' She went off to find it.

Franklin said, 'It's nothing personal, Gordon.'

'You and your friends stitched me up.' Gordon's head was swimming with plans. There was a cyber café in town. He could e-mail *fuck you* messages, with copies to the relevant vice president; messages he should have sent months before. He played it cool under Franklin's gaze, knowing that nobody would support a loser.

Honey came back and the couple left. They headed downhill.

Madame brought a carafe and two fresh glasses. She said, 'There is time before the bus. Perhaps we share a few minutes together?'

Gordon was suddenly aware of the breeze playing with the dust at the side of the road, and that she wore no perfume. He had known Madame as a lanky teenager; a husband had come and gone. There were no children. Now he found himself close to a woman who was enjoying an unexpected break in an otherwise busy day, and chose to share it with him. It salved something of his failures.

'I could think of nothing nicer.' He had a fleeting out-of-body sensation, of looking down on the two of them from the trellising, and seeing a new future.

They drank slowly. It reminded him of the early days in his marriage, of sitting in a good restaurant enjoying an aperitif, anticipating the meal and the lovemaking still to come. He felt he had to say something. 'Your English is very good.' He could have kicked himself for being so inane.

She leaned forward on her elbows. He had to drag his eyes away from her cleavage. 'Very nice,' he said; and added after a short pause, 'The wine.'

She smiled and explained. 'I work in London and New York to learn the hotel trade. Then papa dies.'

'I am sorry.' He said it into her eyes. They were brown, and flecked with yellow.

The smile was still there, and all the more beautiful because of its sadness. 'He is still alive to me in his dream.'

'Dream?' He felt he should do better.

She indicated a derelict extension to the auberge. She drank and became impatient with herself. 'I make myself sad. Worse, I make you sad, and that is bad for custom.' She leaned forward again, closer this time. His head was drawn to hers.

'There is only you and me,' she said. His body began to tingle. 'I tell you something sad, now you must tell me.' There was moisture in her eyes and the turndown of her mouth was pronounced.

The need for sympathy was an ache in him. 'Last year, we were taken over by an American multi-national. There were problems with the takeover and they sent Franklin. He was supposed to be liaison, all the time he was angling for my job. And if that's not bad enough, he followed me here to gloat about it.' Gordon's voice trembled. 'I'll bet everybody at the hotel knows I'm for the chop.'

'Always you think of yourself.'

He started. She said, 'All these years you walk to Auberge Zilla, once, twice on your holiday. Always you walk; always you are on your own. And always you are polite. You are sorry that papa is dead, very formal.' The temper-flush burned in her face. She stabbed a finger at the mountains. 'There are the mountains, there is the waterfall. Not once do you go, not once do you dream.'

She jumped to her feet. Her chair went crashing. 'So you lose your job, Mr. Cold, Mr.... Mr.... Dour.' She blinked at her use of that word; it broke her chain of thought.

He stood with his fists clenched tight to his side. 'My bill, please.'

'Five euro.'

He pulled a fold of notes from his hip pocket and pealed off a ten. She snatched the note from him and crushed it out of sight. He wanted to walk away, he knew he would never be back, but felt challenged to speak. 'I gave the whole of my working life to one firm, and now I'm redundant, finished.'

She laughed. 'How could your life be over, you have never lived?'

'Nonsense! I have a wife, children. I swam for my country in the Olympics.'

'And you are sad.'

Something caught in his throat like a fishbone. He walked away.

The bus wasn't due until three thirty. Gordon stood across the road from the auberge, resolutely looking at where the blue of the sky melted into the hills, and tried to think of nothing. The cicada started up again.

He heard the slap of her sandals on the asphalt and stiffened. 'Monsieur?'

'Yes?'

'I show you something.'

He turned, pretending reluctance. She stared up at him until he had to smile back. He noticed that her dress was fresh on, that her hair was darker as if roughdried with a towel.

She placed a hand on her breasts. *'Je suis Elle.'*

'Gordon.'

'Gor-don.' She rolled the word off her lips. He longed to kiss them. She linked his arm in hers. 'I show, I tell. It might help.'

He feigned reluctance, and his eyes were for her, not the auberge or the stalled extension she talked about. He wondered why he had never before seen Madame as a desirable woman. Her beauty was in her maturity, the lack of gaucheness, the dark hair tied back in a single French plait, the casual disdain of the first grey hairs. He slipped his arm round her waist.

Her father had tried to turn a remote auberge into a select hotel, but funds were limited and costs spiralled. In the end all he had left was a half-completed building, a large bank overdraft and his life assurance policy.

Gordon's mind focused in on what he was hearing. 'Oh my, God!'

She was breathless with passion. 'Papa always said "Dream, and live that dream on your own terms."'

'I am sorry.' He felt sickened.

She spat the words at him. 'Don't be. This year, next, ten thousand euros more, and the Auberge Zilla will be full every night, every room taken.'

'That's great.' He said, and jumped at the rattle of a diesel engine and the hiss of air brakes. He cursed the bus's timing as they separated. People got off the bus. She turned away and wiped at tears with a tissue.

He asked, 'Are you all right? Can I get you anything?'

She shook her head and took two deep breaths, the first one quavered. She said, her voice was almost normal again, 'Friday, I look out for you.'

Friday was days away. He stood close and she leaned into him.

'Why?' he asked, meaning. 'Why so long?'

'Friday you walk to the waterfall. There you dream your dream, and I collect you.'

'I can't ask you to do that.'

'You do not ask.'

He looked at the rows of tables set out under the arboretum and could see himself working there. He was shocked at what he was thinking. A new beginning, and things uncomplicated.

★

On Friday Gordon came in his car, racing-blue and family comfortable. Elle showed him the old smugglers path behind the auberge.

The path was easy to follow as it rose in slopes and steps to match the fall of water. There was a spring in his step, and he walked with his head thrown back, enjoying the freshness of the air and the shimmer of heat haze on the far mountains. The warmth reminded him of Elle's touch. Eventually a noise grew; faint at first then louder, always constant, like the sound of a motorway heavy with traffic. He turned a corner and, suddenly, the waterfall filled his world. Seventy feet of fury crashing into a pool that boiled in the maelstrom.

His shirt was sodden with sweat. He pealed it off and took the final steps into a pudding bowl valley. Running the rim of the valley was the road, he could see no car waiting, but waved anyway. Only the bushes caught in the vortex of air from the waterfall waved back, the only call that of the cicada. He was tempted to pound his chest and do a Tarzan call. Then he stopped, surprised. Elle was there, stretched out on the ground, with a large sunbonnet over her eyes. Other than that she was naked.

'Elle…? Madame…? Ah…' There was no response. He wished that she was awake and teasing him with pretend modesty, because startling her or doing the wrong thing, could spoil everything. He wondered if he should wait, but anxiety, fear for her that she might be hurt or in a coma, drew him on.

He called, louder this time, raising his voice against the roar of the waterfall. 'Madame,' and again got no response. He went closer, hesitant, ready to stammer explanations, until he was looking down at her gold-tanned body. His shadow touched her face and her eyes opened. He turned and stepped away.

'I slept,' she said, still drowsy.

He kept his eyes fixed on the far rock and stretched his ears for the sound of her dressing. He thought she moved. It was more an awareness than a noise.

'I slept,' she said, again. Her voice sounded brighter, more normal, and he stopped worrying about her. 'And I make you…' She hesitated, searching her mind for the right word. 'England is always cold. Here we have the sun and pretend we do not see the body.'

He was frightened of a car coming while she was dressing. He wished they could melt into each other. He almost smiled at his own confusion.

'You *may* turn if you *can*. Is that not correct grammar?'

He turned. Her dress still lay on the ground, the centre depressed where her head had lain. She was sitting, holding the hat over her hips and laughing at him. The breeze caught the rim of the hat and lifted it, revealing the darkness underneath. He dragged his gaze away.

'Oh you English.'

'Irish.'

'English, Irish.' She made a face and pointed at the ground beside her. 'Sit.'

He didn't dare refuse. He didn't dare obey.

Her voice was gentle, understanding. 'I embarrass you.'

He had to clear his throat. 'I think...' He cleared it again. 'I believe the Chinese do the same, not see the body I mean. There are so many of them.'

She patted the ground. 'So you will sit?'

'Er... yes.' Now it was harder to refuse than obey. He sat and continued to stare at the rock and was furious with himself. It was years, a lifetime of marriage, since he had been in the market, and he didn't know the right moves. He was aware of her watching him.

She said, 'Every year, on papa's anniversary, I promise I come, and every year...' Her hand touched his leg, and he caught his breath. 'Thank you for making me do this.' The hand remained. 'I dream of you. Better that than of my last visit; *les sapeur-pompier* bringing papa's body up the slope and the nosy watchers, always them.'

Gordon went cold and he looked round, half expecting to see papa's ghost watching them.

He put his hand over Elle's and gave it a comforting squeeze. She said, 'Do not be sad, I am not.'

She used the sunbonnet to fan herself; he kept his eyes firmly turned away. She laughed and pushed it into his hand. 'A gentleman would keep a lady cool.' She lay back.

He knew nobody could see them. She was... charming, and she was wearing a perfume, something subtle with the essence of the countryside about it. He swayed the sunbonnet to and fro. 'It's hard not to see your body.' He felt good about that little joke, especially when she smiled.

Her eyes closed and her breathing deepened. He began to enjoy himself. His trousers were hot and uncomfortable in the sun; he slipped them off, but kept his briefs on. They were new and specially chosen for the day. He was embarrassed for his legs; they were pasty-white against her tanned body.

Elle's eyes popped open, her hand came up and drew him down. He shifted some of his weight onto her as they kissed, and wished he had dared take his briefs off as well. He nudged at her legs with his knee. She stopped kissing him. He thought he might have rushed things, but it was too late to back off. He gyrated his hips against hers.

She pushed him away and stood up.

He said, 'Sorry.' It seemed to be the right thing to say, and pulled his hip muscles tight to flatten his stomach.

She ran on tiptoe to the edge of the grassy ledge, and stretched hands and body upwards, and laughed at him as he stumbled out of his briefs. He kicked them clear. She arched a perfect swallow dive into the void.

He ran to the edge. It was a sheer drop into the pool at that point.

She was gone, he couldn't see her, he was sure she was never coming up. He was counting. 'Twelve, thirteen.' He would have to rescue the stupid woman. 'Sixteen, seventeen.' He would go at twenty. He counted more slowly. 'Eighteen, nineteen...'

She came bursting into the air, glorious in her water-slicked tan. She trod water and shook her hair free of the plait. 'Come,' she called.

He glanced up, no cars were passing. Nobody to see and tell his wife. That thought made him hesitate. Elle pushed herself onto her back and floated in the boil of water. Her darkness drew him. He flung himself after her.

The cold of the water fire-flashed through his body, his heartbeat faltered.

Stopped.

Started again.

He struck for the surface, breathed too soon, choked and dove into the fresh air. Two quick strokes took him to the side. The rock was unclimbable; he held on with his fingertips and blessed the pound of his heart.

She raced to him and supported him. 'Gordon, I forget, the water is from the snows.' Her eyes were glowing.

'It's bloody dangerous.' He choked again.

The cold became bearable. She kissed his lips, sliced water in a quick crawl and was halfway up the slope when he was still in the water. A bus came; she ducked into a tumble of water-smoothed rock. The people on the bus cheered and waved. He waved back and tried his Tarzan call. She popped out again when they had gone. 'I think you go mad.'

He could hardly make up his mind which part of her he wanted first. He spread-eagled her against a rock. 'Not here, not here,' she kept saying; yet her nails scored furrows of delight down his back. He tried to enter her where she stood. She brought her legs together, and held them against his push. He backed off, frustrated.

Again she led, as they climbed the side of the waterfall together. The coldness came off the water like a frost; he shivered and clung to her hips for warmth.

He didn't see the narrow ledge until she stepped onto it. He followed her, sidling along a slimy wall, until the fall of water was only a nod away, pounding their ears, making the world vibrate beneath their feet.

They stopped in a niche between rock and water. She turned into him, her heels teetering over the abyss. Her eyes gleamed a dull gold as she pressed hip against hip and leaned back.

He shouted, 'Don't!'

Her head went into the falling water, disappeared as if cut off at the neck. She was forced away from him in a backward dive. His balance gone, all he had left was the old instincts: head ducked into shoulders, arms extended. For a moment, he felt as if he was being pounded to pieces in a grey hell then he was underwater. This time he was expecting the cold and it was bearable.

A rock like a giant football stood alone in the bottom of the pool. He was heading straight for it. A quick twist of the body, a solid shove with the palm of his hand, and he was clear and striking for the surface.

She was waiting for him. They kissed and slid underwater, air bubbled when they tried to laugh. She led him to a flat rock lying half in and half out of the water, its surface covered with thick lichen. She rolled onto it and held her arms out for him.

'This is for the waterfall,' he said, and smeared lichen over her breasts.

She squealed and fought back, he smeared lower.

She grabbed her own handful of lichen and ground it against his penis. 'There, who take that now?'

They fought and smeared, tumbled into the water and came up holding hands, and looking into the depth of each other's eyes. She lay down on the rock again. He was gentle with her, she was gentle back. He wiped a stray hair from her face.

The sun was warm on his back, cars passed above.

After an hour of effort and somnolent bliss, they stretched off the rock and climbed the slope to their clothes. He held her in a loose embrace; she rested her head on his chest.

'That was good,' she said.

'Good? It was... You were like a dream.'

'Then I am happy for you, Monsieur.'

She slipped her dress over her head and searched with her toes for her sandals. His embarrassment at his nakedness returned, and he turned away.

She curled her legs into the grass while he dressed. She said, 'Today I give you your dream.'

'The auberge?'

She smiled. 'Next year, God willing. The cents add up.'

He knelt facing her, their bodies mingled naturally.

He said, 'Ten thousand euros, that's only six thousand pounds. Surely the bank?'

'Never! I gave papa my word, no debt.'

An icy coldness touched Gordon; she must have known that her father was going to kill himself. 'Couldn't you...?' He stopped, he was only guessing. It was too late anyway.

She looked sad and shook her head. 'He would do it. He was very brave when it didn't work the first time.'

Gordon's heart was beating hard against hers. Six thousand pounds, the holiday was costing that and more. 'If I lent…' he hesitated to commit himself further at that point.

'Do you think I am a whore? Do you think I give my body for money?'

She was stiff now, turned in on herself; he eased his grasp. She stared into the waterfall and began to shake.

'Oh Elle.' He soothed her and wished he could tell her of their future together. Then was not the time. 'Give then, me helping your dream along. For your papa's sake.'

She wrapped herself around him and held on tight. He wondered which of the shuttered rooms they would use. The bed was bound to be iron and erotically creaky.

'There, there,' he said. The shaking stopped. She wiped away the tears and kissed him.

In ten minutes they were back at the auberge.

The little car park was nearly full. A couple walked past them as they sat in the car, they were young-middle aged and obviously impressed.

Gordon said, 'Things are looking up.'

'Yes, it is a good day.'

She went to touch his face. She looked round, there was no one in sight, and still she hesitated. She looked sad. 'Do not forget me.'

'How could I?' He brought her hand to his lips. 'Forever and ever, until death us do part.'

She looked puzzled, her smile was uncertain.

Gordon told himself not to mix money with the proposal. He kept a spare cheque in his wallet for emergencies. He produced it, and what could have been an awkward moment became something special.

There was a cry and the sound of glass breaking.

Her fingers dug into him until her nails marked his skin. 'That girl! I must go.'

'Wait!'

The second cry was more sustained. She joined hands in supplication. 'A minute,' and disappeared round the corner.

He went into the auberge and stood at the unstaffed reception desk, ready to answer the phone or help if needed. His knew his holiday French needed upgrading; he would start on that that night. Yet, with Elle no longer with him, he became unsure of his feelings and wondered what his wife would say if she ever found that particular bank statement. He could drive off. Elle would never see him again and wasn't likely to follow.

He shook his head. Nothing remained of his marriage except habit.

He remembered the water, and the flat rock, and Elle giving her body to him. He had been so cold by that time, so up and down – he smiled as he remembered her words – that he hardened slowly. She thought he was playing with her and pleaded for release.

The second time they made love she mounted him, determined on revenge. 'There you see. What do you think when I hold it back? It hurts, doesn't it?' It was the most delicious agony he had ever experienced.

He straightened his shoulders. 'Hell roast it, the woman's entitled to her dream. And so am I.'

He wrote the cheque and wrote it clearly, Madame Elle Defarge £6,000, and made the signature the best he had done in years. He wanted the whole thing to be a creation, to be seen as deliberate, not as an act of impulse. A final dot at the signature and he was finished.

And suddenly deflated, for there was no one to share his moment of decision. He could leave the cheque on the desk, but it could blow away or be stolen, or he could go looking for her and present it formally. She wouldn't like that.

The window was open, and he became aware of other people and of Elle's melodic laughter. If he could catch her eye? He moved towards the window and stopped suddenly when her companion spoke. It was a man's voice, American, and unmistakably Franklin's.

'Big things are happening. I have to be at the office first thing Monday.'

She said, 'Big things for you, I hope?'

Gordon's mind was screaming. 'My job.'

'Yeah.'

'Then this is your last day?'

Gordon thought. 'How can she pretend to be so interested in

him?'

Franklin said, 'It's my last chance to walk these hills and breathe God's good air.'

Her voice dropped, Gordon had to lean closer to hear the words. 'Then you must follow the stream to the waterfall. There you will dream your dream.'

Gordon was stunned.

Franklin said, 'How do I get down again, whistle a cab?'

She said, 'I collect you.'

The only part Gordon could move was his hands. He took the cheque between thumbs and forefingers ready to tear it up. He mouthed the word. 'Bitch! Bitch! Bitch!'

'Little lady, I couldn't ask you to do that.'

'You do not ask.'

She talked on, coaxing Franklin. Gordon stood and listened, and remembered.

He remembered how her eyes had glowed when he surfaced from the first plunge into the pool, the time his heart nearly stopped. Remembered the same glow when he fell from the waterfall and almost hit the sunken boulder; only luck and fitness had saved him.

And the dead skinny-dippers?

He went cold, he went hot. He found an old bench and sat because his knees wouldn't lock. He was shocked at her leading him on. 'And I nearly fell for it,' he said to the cheque. As soon as he could he would rush to the door. Franklin had to be warned. Franklin had lung problems.

But why? Why did she risk lives? Kill?

Realisation jerked him back like a blow to the face. Her father *was very brave when it didn't work the first time*. The old man had thrown himself into the water and, when that failed, took a header from the ledge onto the football rock. Elle had to be there to know what happened.

It burst out of him. 'Dear God Almighty.'

That was her compromise, her terms for saving the auberge, collecting off those who survived.

His knees were working again, he snapped to his feet.

He remembered something else. Elle in his arms, and he drawing the essence out of her. Her eyes glowed then. He was that good? He shook his head in wonder.

Do not forget me. She meant that.

Gordon left the cheque on the counter and walked to his car.

PLAYING WITH FIRE

They were waiting for me when I arrived in the yard at nine thirty sharp, as ordered. Training Officer, Sergeant Vera Barlow - very like her father, they said: big, thick and awkward. And beside her, Luke, who was stripped down to his trousers: no shirt, boots or socks. Luke had no second name or rank that I ever heard off. He was all ribs and collarbone, and his stomach was so concave his belt buckle threatened constantly to touch his spine.

I snapped to attention before them. 'Constable Crawford reporting as ordered, sergeant.'

Vera Barlow's voice was cold, 'In this command you have to earn the right to call yourself *constable*.'

'Probation Student Crawford J, 416, reporting as ordered, *Sergeant*!' I bawled into her face.

We stood nose to nose. I had enough sense not to stare directly into her eyes, but focussed on the tip of her nose. I could hear the hoarse screams of the drill instructors on the square and feel the thrum of engines from the Land rovers in the next bay.

'Strip. Same as Luke,' barked Vera.

I hesitated, but Vera gave no explanation. She and Luke just waited on me. I stripped off, put everything neatly against the wall and turned to face them again, saw an attack movement and managed to partly block Luke's kick. His foot still slammed into my ribs and drove me to the ground. Then he stepped back and waited for my reaction. An instant audience poured out of the surrounding buildings.

I got up slowly, rubbing my chest. 'What was that for?' I asked Vera. She at least looked like he was enjoying herself.

Luke came at me again, a high-pitched yell screaming from his throat. He stopped, pivoted and lashed back at my face with his heel. I caught his foot mid-swing, and pushed it up and away from me. He crashed to the ground head and shoulders first. Still holding his foot, I turned sideways to him, slammed my heel into his lower stomach and jumped back and clear. Luke curled up into a ball and lay still.

Vera ran forward. 'That was deliberate,' she accused.

My eyes remained fixed on Luke who lay where he had fallen. I kept my tone light. 'He hit me. I hit him.' A shrug. 'Fair's fair.'

Vera laughed, genuinely amused. 'No. I mean you really can do unarmed combat. How come?'

Luke rolled on to his knees and knelt back on his ankles, listening. He kept his hands pressed to his stomach; if he was in pain he didn't show it.

My answer was evasive. 'I've been around.'

Luke got to his feet, an act of sheer determination, and took his usual place slightly behind Vera. I tensed as he approached, but there was no light in his eyes and I relaxed again.

Vera checked Luke's eyes for herself then turned back to me. 'And you became an expert?'

I hesitated. Belt colours and dans, and chasing an unrequited love through a series of sports, did not form part of my CV when I applied to join the police, and would take a lot of explaining. 'Only so far,' I compromised. 'And even then I'm out of practice.'

Her ready humour bubbled to the surface. 'Well, me boyo. The Commander asked Luke to teach you all he knows about unarmed combat. I be thinkin' you had better be teachin' Luke first.'

I had vaguely assumed that everyone went through the same course. I was surprised that this was to be a series of lessons put on for me personally. The thought pleased me, even if I was cynical enough to wonder if the Commander hoped Luke would kill me in the process. The Commander was my father from his first marriage, and Vera his long-time partner. We didn't get on, not with my mother dying of cancer and me being raised by grandparents, and no contact. Not a letter, nor a birthday card, nor a Christmas present, in twenty years.

Vera gave Luke a friendly thump on the chest; no one else dared touch him. 'Thirty minutes, five days a week, Luke. No more.' Then she added, more seriously. 'Bend him a little, but don't break him.'

'Thank you very much,' I said dryly.

She walked off, and left me toe to toe with our not so friendly neighbourhood lunatic. Luke went *on guard*.

I took a long breath to steady myself, and tried to take control of the situation immediately. 'Good. Good,' I said. 'Your position is not bad. Now, try it with your right foot slightly more forward.' I tapped the required spot with my toe rather than risk touching him. 'And your hands wider apart, and higher.'

Luke was good. Someone had given him the basic idea. Application and a natural poise had made him an expert in unarmed combat. He only had to be shown a thing once. I demonstrated, he copied. I went wrong and put us both wrong, and apologized for being out of practice. It was like talking to a brick wall - there was no reaction.

After a time, I relaxed sufficiently to allow us gentle swings and prods at each other. Right in the middle of an attack Luke stopped, turned and walked away. The clock in his head said that the thirty minutes was up.

Luke drained me of my knowledge in less than two weeks. That part of my training ended as brutally as it had begun. In the middle of a lesson, Luke suddenly felt he had sufficient knowledge to beat me, and turned a practice round into a serious battle.

He attacked as the first time, and I could only block the blow. There was never any chance of catching his foot. His attacks were fierce, mostly kicks and punches. Light of build, he realized he was at a disadvantage against me on the ground, and kept his distance. I could do little but defend and block and, at each attack, give ground in a large circle to avoid being trapped against a wall or in a corner. We both took several knocks to head, arms and body.

It seemed like hours, but it lasted only five long minutes, until our time was up and Luke stepped back and turned away. Our regular audience clapped and cheered as we walked off.

The next morning he came out carrying two Bowie knives. These broad bladed knives were honed to a terrifying sharpness; the hooked points were needle sharp. He dropped one at my feet. We had a larger audience than usual. Luke's eyes had a gleam of anticipation.

I picked up the knife and went on guard. My lessons at the training depot had included a course on dealing with an armed attacker, and to know the defensive movements you had to know the offensive thrusts. But knife fighting was Luke's specialty, his joy. I was well outclassed from the very start.

He advanced in a threatening manner. I grasped the knife and braced myself. He went through the attack sequence slowly, assessing my reactions, and made me repeat them against him while he showed me the counter moves. Even on that first day the tempo increased, strike and guard, strike and guard, without let up.

In an instant Luke changed. His eyes glowed bright and he came at me; I had no time to feel fear. He swept through my defensive moves, smashed me to the ground, and twisted my knife arm inwards and upwards until the skin burned as the blade cut into my neck.

I was helpless, and lay with my knife embedded near my jugular vein. Luke's eyes were burning coals and his face was alive with joy - the joy of the kill. There was silence in the yard. No one dared move.

Out of the blue, a thought. I whispered, 'Bend him a little, but don't break him.'

The light died in Luke's eyes. He got up and walked away.

The half hour was up.

MADONNA OF THE FALL

The parquet flooring emphasised the halt in Michael's step. A chair stood in front of the desk. He put a hand on its high back to ease the ache in his leg, and looked steadily past the bishop's head to a painting of the Madonna of the Rocks. An anguished Christ nailed above the door was the only other wall-hanging in the room.

The bishop waited a time before he asked. 'You are decided, then?'

'Yes, your Grace,' said Michael, his determination hardened by an extended wait on a polystyrene chair, chosen, personally by the bishop he believed, to cause the utmost discomfort.

'And you are willing to throw everything away? On a whim?'

Michael remained silent and kept his gaze steady on the Madonna, he hoped she at least might understand. The painting showed only the Virgin's head and shoulders against a murky grey background. Instead of a veil she wore a white mantilla, and her shoulders were bare. The priests of the diocese called her the Naughty Madonna.

'A retreat,' said the bishop, using the power of his preaching voice to emphasise his total control over Michael's life. 'A month, two, of prayer and meditation on the sanctity of your priestly vows.'

'No,' said Michael. He took his hand off the chair and firmed his back. 'I have tried everything you ordered and I can run no more.' He drew breath; he found he shook like an aspen in the wind. 'I am sorry.' He turned to leave.

The bishop jumped to his feet. He was a sparse man, but the thump he gave the desk echoed loud. 'You walk out of this room, and I will curse you with bell, book and candle.'

Michael felt sick with anger. He said, 'It'll be a bad day when I serve God out of fear.' He walked on, grasped the door handle and looked back. 'Your Grace, will you damn me for being honest?'

The bishop gave the table another thump. 'That's a question you should ask of God. I can only damn you for abandoning your parishioners.' He picked up a computer listing of the priests in the diocese and flicked through it. 'You are no longer Parish Priest of St Joseph's. You are now the curate of St Gabriel's.'

'Where?' asked Michael, knowing he should wish the bishop a good day and leave. He remembered hearing vaguely about St Gabriel's. It was ….

'Buried in the mountains,' said the bishop. He looked calmer. 'The Parish Priest, Fr Toner, had a heart attack last night.

He threw the listing of priests at Michael; the fan of sheets unravelled in flight and fell well short. 'You tell me, who is available to replace you at Saint Josephs?'

'I'm sorry,' repeated Michael, and again heard his own limping footsteps as he picked up the listing and put it back on the bishop's desk. The delay gave him a chance to harden his heart against St Gabriel's. He remembered hearing of it as a laughing threat among the priests. A remote hamlet that straggled up a long valley, a one priest posting. The parish priest, Fr Toner, had been there pre war. First World War said the wags at assemblies.

'No,' he said.

The bishop glared at him. 'Fr Michael, must I remind you of your sworn duty to serve God? There is nobody else I can send.'

Michael drew a deep breath.

★

The little parochial house was built of semi-dressed lava stone and blue-slate, the church the same, and both needed fresh paint. Jo stood behind Michael, he sensed her cold stare cut through him as the front door creaked open to reveal a housekeeper sagging with age.

The housekeeper wore a wrap-around apron and shapeless slippers. 'Fr Michael?' she asked.

'Yes,' said Michael. He held out his hand. 'I take it you're Olive.'

Olive exchanged a brief handshake, her look shifted to Jo.

Michael said, 'This is my friend, Jo Burnside. She will staying with us for a few days.'

'There's no room for Miss Burnside or anybody else,' said Olive.

'Mrs,' said Jo.

Olive stood on, blocking their entrance into the unlighted hallway; behind her a dull light glowed from the kitchen. 'You'd be more comfortable back home with your husband,' she told Jo.

'He's dead,' said Jo sharply, and her colour heightened.

Michael frowned at Jo for venting her temper on Olive. Usually it was of the explode and go variety, this outburst had lasted for hours. 'The posting's only temporary,' he had assured her. 'A couple of weeks.' Jo chose not to believe him, hence the temper.

Caught between Olive and Jo, he hesitated. Then Jo took a step forward. Once she had done that he knew she wouldn't run.

'I'm Fr Toner's sister,' said Olive, giving way. 'And this is a small house, Father,' she continued, as if there was a cause and effect between the two announcements. She flung one door open. 'Sitting room, that's for your friend. There's an old bed settee. It'll do rightly.

'Kitchen,' she said, indicating the second door, then creaked up the stairs before them, all the time muttering under her breath about hoity-toity people.

'What did you say?' asked Michael, annoyed.

'I never spoke,' said Olive and pointed from the top step. 'Your room, Father, my room, bathroom.'

'I'll sleep in the sitting room,' said Michael, taking comfort from Jo's hand on his back, even if it did tremble.

'You can't. The only telephones are here and in the kitchen,' said Olive. That announcement seemed to please her.

'The sitting room's fine,' said Jo.

Olive pushed past them and shuffled heavily downstairs, muttering about visitors and being put out, and it was all right for them as had the health.

Michael and Jo made a face at each other and went into the bedroom. It was full of another man's presence: a priestly vest and collar hung behind the door, and the dressing table drawers were warped with age allowing over-washed clothes to gape out. An air of polished hopelessness and prayer hung in the air.

Jo shuddered. 'I'd hate this room.'

'The whole thing's impossible,' he agreed. They sat on the edge of the bed and he took her hand.

She grasped his back. 'Sorry, I've been a bitch.'

He said, 'I should have said no. But I felt God….'

'What would he know about it? He's another man.'

Michael relaxed, once Jo voiced heresy he knew things would be okay.

Without being asked, she began to massage the damaged muscles in his leg; their knot of annoyance at the long car journey started to ease. They planned as she worked. Tomorrow he would see about renting a house for her, something nearby, while she went looking for a job at the local hospital.

'No hopeless patients,' he told her.

'And no despairing widows,' she replied.

Olive called up the stairs. 'Father, there's tea made, and Granny Ingram could do with the Blessing of the Sick.'

Jo stood up and brushed her skirt straight. Everything she wore tended to be square and official looking. *Matronly*, Michael called it when he wanted to annoy.

He checked the Ingram address on the map while he gulped down a cup of tea, the road to the Ingram house wound through steep contours. He shook his head, a city man born and bred he knew he was bound to get lost. 'I'll keep asking,' he said.

Olive gave a *tech* of impatience. She led him into the hallway, muttering to herself about people who couldn't find their way on a straight road, and pointed at an oil painting of mountains folding in on a winding country road. 'There,' she said, and put her finger to a red dot on one of the bends. 'That's the house.'

Michael laughed. 'I think even I could find that,' and ran for his car.

Michael was late back from the Ingrams. Granny Ingram was used to having little turns. 'It's God's will,' she said, and remained calm even with the pallor of death on her cheeks. Michael, as usual, felt humbled in the face of such faith and led the family in prayer. Afterwards, they took him into the family room for a subdued supper. He left when he caught himself starting to yawn.

'You're in pain,' said Jo first thing when he got back to the parochial house. He nodded. Pain when he was over-tired, and the limp, was a lasting legacy from a car crash ten years before. Jo was working Triage the night they brought him into the hospital to die. When she couldn't think of any further medical procedures to keep him alive, she accused him of being gutless and not fighting and, anyway, how could she and her Paul get married if he wasn't there to perform the ceremony. He must have heard her shouting because his blood pressure popped up and he never looked back.

His and Jo's suppers were in the oven, going dry. Olive had eaten earlier and was crocheting at the table while she watched television. The anthracite stove made the kitchen oppressive with heat, and the old square table and worktops left little room for moving around or stretching one's feet.

Michael looked at Jo, who glared back grim-faced. Michael reckoned Olive could hear a pin drop so they ate in silence. The meal had been lovely, braised steak done in some tasty sauce; he said nothing about the Ingrams feeding him. The apple pie that followed was delicious, though they passed on the custard, which had gone solid.

'Any news of Fr Toner?' asked Michael, to fill a silence.

'As well as can be expected,' said Olive, and went off to lock up.

'It wasn't too bad,' said Jo, loyally, before he could ask about her evening. 'I think she's worried what will happen to her if Fr Toner dies.'

It hadn't occurred to him. He felt guilty as he picked up the phone.

Jo asked. 'What are you doing?'

'Ringing the bishop.'

She shook her head at his stupidity. 'You'll get nobody there this time of night.'

'I can leave a message on the answering machine,' he said. It seemed safer than having to talk to the bishop direct, or to the diocesan secretary who had been so self-righteously disapproving that morning.

The bishop answered on the second ring. His 'Yes?' was testy.

Michael gulped and identified himself. Jo made a face and slipped away.

The bishop said, 'This had better be important, Father. My secretary is now Parish Priest of your former parish, and I have to do my own paperwork until I can get a nun. If I can get a nun.' He cut off Michael's stuttered explanation as to why he was calling. 'That stupid woman. Did she really think I'd forget her? Put her on.'

Michael called Olive to the phone. The bishop was by turn friendly-cross and cajoling with her. Olive denied ever having any doubts that he would forget to look after her, and wiped furiously at her eyes with a crumpled tissue from her apron pocket.

Michael heard Jo in the bathroom then the pad of her feet on the steps. As soon as he could, he wished Olive goodnight and joined Jo in the sitting room. The bed was already made up, and she had an Ibuprofen and a glass of water waiting for him.

Jo pointed at the bed. 'If I don't get the knots out of that leg you won't get a wink of sleep tonight.'

Michael undressed and lay on the bed while she massaged his back and leg. Jo told him to turn over. She was wearing light blue pyjamas with minute flowers dotted over the jacket. He smiled up at her as she worked; she pretended not to notice. Finally she looked fearfully at the door. 'She'll hear.'

Jo had a well-rounded figure and a neat waist. Michael took her pyjamas off, the better to admire her. 'We'll be quiet,' he said.

'You can't stay.'

'I won't.'

A woodpecker tap-tapped on the tree above Michael's head. He wanted it to go away, it was disturbing him and the sunlight glared in his eyes. He came awake. Olive stood in the doorway with light from the hall spilling in around her. He stayed still hoping she wouldn't see him.

There was a pause then she said, 'Father, it's ten past seven.'

'Right,' said Michael.

There was nothing else he could say. He was in Jo's bed with Jo lying half on top of him. The bedclothes had fallen down far enough to show that they were naked.

'If you're awake,' said Olive, going out and closing the door behind her.

'Fuck!' said Michael into the darkness before slithering out from beneath Jo and finding his clothes. He trembled as he dressed, wondering if Olive would allow him to say the seven-thirty Mass. She was probably on the phone to the bishop as it was. He thanked God

Jo was a heavy sleeper, and he didn't have to face her wrath as well.

He let himself into the hall. The light was on in the kitchen, but he sensed the house was empty. Through a side-window he saw lights come on in the church. He raced upstairs to freshen up then went across to the church. The sacristy was unlocked and the vestments of the day were laid out for him. A young alter boy, Shane Ingram, waited to serve Mass.

'Ingram?' said Michael. 'Are there many Ingrams in the valley?'

'Lots and lots. It's great at Christmas,' said Shane.

Michael laughed and they processed out onto the altar. Olive came up for communion. She was one of the old sort who preferred to receive the sacrament orally instead of into her hand. That meant she had to tilt her head up, but she kept her eyes closed so that they wouldn't meet Michael's.

After Mass, Michael usually sat on and read his Office. That day he thought it best to go straight to the parochial house rather than leave Jo to face Olive on her own. He found the two women getting in each other's way as they made breakfast, Olive muttering all the while about a body not getting peace in their own home. He sat well back and started into his Office until Jo gave him his usual cereal and toast. Olive offered him a share of her bacon and eggs. He didn't dare say yes.

'Did she say anything to you?' he whispered under cover of the sizzling bacon.

'What about?' asked Jo.

He thought he'd better tell her later.

Olive sat across from them. There was a silence. Jo's look said that she had been trying to make conversation with Olive while he was in the church, and now it was his turn. He said, 'Olive, I need a list of people I should visit.'

She nodded her agreement, and he didn't know what else to say after that. The scrape of cutlery on bowls and plates made things worse. Jo went off to get herself ready to go out.

Michael cleared his throat nervously. 'About last night….'

Olive interrupted him. 'Granny Ingram is a bit better this morning; they rang.' She bustled off and fetched the map; he was surprised at her speed across the floor. 'It's my arthritic knees,' she said, as if reading his thoughts. 'They won't make the climb.'

He shrugged inwardly, relieved. Olive didn't want to know.

The map was old and beautifully drawn freehand, and many of the houses had family names marked against them. 'There's not many changes,' she told him and began to explain the layout of the parish. She also gave him a list, written in a shaky hand, of all the current sick or infirm he should visit. He groaned. They all seemed to live in the furthest-flung regions of the parish. He took the list and map into the hall, to the painting he had used the previous evening to find Granny Ingram's house, and from there went to other, similar, paintings, identifying parishioners houses. Olive went with him and was helpful in a neutral sort of way. One painting didn't tie in with the map. Like the rest it was of the countryside, but a bleaker terrain. He looked at Olive, puzzled.

'That's Norway in the spring,' she said.

For the first time he looked at the pictures as works of art. They were executed with the confident strokes of a strong personality with a great gift, and all of them were unsigned.

'Who was the artist?'

'A friend,' said Olive. 'Now if you will excuse me, Father, I have to get ready.'

Michael nodded. Jo was going to the hospital to look for a job and intended to take Olive with her to see her brother. In the meantime he thought he should use Fr Toner's car to do some house calls in Granny Ingram's direction, it would give him a chance to call with the old woman again. He couldn't understand how Fr Toner hadn't suffered a heart attack years before, he must never have had a spare moment to himself. If it was him…. He stopped himself thinking that way, St Gabriel's would never be his to organise.

Michael had a great day calling with people and getting to know them. Not that they gave much away, not valley people, not on first acquaintance, though he did learn that Granny Ingram's maiden name was Kennedy. She told him. 'Throw a stone in the valley, and if you don't hit an Ingram you'll hit a Kennedy.'

A sharp nosey woman, one of the Kennedys from the Coll, with a voice that carried half across the parish, stopped Michael near the end of his sick calls. 'You're living in the parochial house I hear?'

'Yes,' said Michael. 'And you'd be…?'

'You and your sister?' she said, ignoring his outstretched hand.

'Not sister,' he said and walked on, annoyed and embarrassed. She was the sort of old bitch who would take it upon herself to phone the bishop. He decided not to unpack too many of his things and, in the meantime, he and Jo should sort something out between them.

He went back to the parochial house to find a note on hospital paper from Jo saying she was working and would be back about six. Olive was busying herself about the kitchen, putting a good shine on the linoleum with her slippers as she bustled about preparing roast beef and three vegetables for their dinner. He thought it smelled pretty good.

'You'd think a body would notice,' Olive muttered to herself as she got him a cup of tea without being asked.

Michael pulled himself out of his own worries. Notice what? he thought. Olive was wearing a shapeless multi-red coloured dress with a thin belt round her broad waist. His granny was the last woman he remembered dressed like that, and he wondered if somebody made them to order or if Olive bought them in antique shops.

'Your hair,' he said, catching on at last. Olive's hair had been permed and given a blue rinse. 'What have you done to it? It's gorgeous.'

Olive hand-freshed her perm and looked pleased. 'Mrs Jo suggested it,' she said. 'They asked her to stay on when one of the Kennedys from Dark Water, died.'

'Who?' asked Michael, half on his feet. He felt he should have been called for.

Olive waved him down impatiently. 'The Dark Water Kennedys are Methodists, but they'll expect you to call. Mrs Jo got blood on that lovely blouse of hers and I had a while to wait for the bus.'

Michael had to blink and think, before he realised they were back to getting the hair done. He sniffed the meat slowly roasting in the oven, the roast potatoes had just gone in, and decided to leave calling until later. Olive produced heated apple pie and put it before him. 'Don't let it spoil your dinner,' she warned. He patted his stomach and thought weight could be a problem if he lived with Olive too long. She walked off muttering about never fattening a thoroughbred.

Michael finished his tea and started on his evening Office; his eyes drifted closed.

Jo arrived home. He knew trouble lay ahead from the bang of the door and the slap of the handbag on the floor. She came into the kitchen, her face flushed and her hair untidy. She had been crying, and a little tick of pressure on her temple worked overtime.

She glared at Michael sitting with his breviary and a fresh cup of tea handy. 'I see you'd a great day anyway, meeting people and getting well fed. What about us?'

'What?'

'Us. You and me.'

Michael's mouth opened and shut. He had forgotten about seeing the estate agent. 'I've been busy,' he said, weakly.

Jo pulled at the top of her blouse. 'This house stinks of heat. Do you ever open a window, woman?' She struggled with the sash window, it slid up and the curtains swirled into the room. Michael sat on, stayed still.

'Dinner's about ready,' said Olive, tight lipped. She had the roast tins and steaming pots circling three plates.

'I don't want anything.'

'Now you've got to eat, Mrs Jo. I cooked it special for you.' Olive held out a slice of beef on a fork for Jo to taste. Jo didn't take it and Olive pushed it closer to her face.

Jo jerked her head back and swatted at the hand. 'I told you, I don't want anything!' The fork and the knife landed on the floor.

Olive bristled, but moved back from the threat of Jo's temper.

'Go easy,' said Michael.

'Easy,' Jo spat. 'You lied to me. "Two weeks in St Gabriel's" you said. Two weeks? Ha! That old man is suffering from congenital heart failure. He'll never be back.' She stood over Michael. 'And what have you done today?'

He closed his breviary and tried to take her hand.

She snatched it out of his reach. 'I've given up everything for you. My home, my job, friends I've had a lifetime will never speak to me again. And all you had to do was go talk to an estate agent.'

She went out and slammed the door behind her.

Olive brushed Michael away when he went to help. She picked the fork and meat off the floor and fiddled with the rest of the meal, muttering all the while about never before being so insulted in her own house and what was the world coming to?

He told Olive to hold dinner for a while and followed Jo into the sitting room. She sat quietly on the settee, her head bowed, looking at something held in her cupped hands. He knew what it was without having to ask. It was a photograph of her daughter, little Philippa, who would forever be eight months old. He sat down beside Jo and let his thigh touch hers, a gentle pressure to let her know that he was there, and waited.

She said, 'Brian Kennedy's children came right into the ward when I was clearing up. They burst right in on me. They were gorgeous kids. Young Brian, he's only three, he didn't understand, but Anthea did, she's older, and they had cards for him from school: *Get well soon, Mr Kennedy. Hope you're feeling better, Mr Kennedy. Love you, daddy*, and I had to throw them out.'

She turned her head towards Michael, but her focus was somewhere through him. 'He wasn't expected to die, you see, but he haemorrhaged, and the children's cards got mixed up with the blood. Their mother was so brave, and she got blood on her as well.'

Jo started to shake. Michael took the photograph of little Philippa from her hands and put it carefully on a side table. Jo continued to stare through him, but he knew she was now seeing that night in Triage when the accident victims turned out to be her husband, Paul, and little Philippa. She started to cry and allowed him to take her in his arms. He knew he should be out doing; there were evening devotions to give and Brian Kennedy's widow to visit. He understood why the Church didn't want married priests, and knew they were still wrong. A family with all its entanglements kept him in the world instead standing somewhere outside in judgement.

The evening slid on. Jo had just roused herself from a deep sleep, and come into the kitchen to make some sort of peace with Olive, when the bishop came steaming into the house unannounced. 'What's all this about that woman living here?' And Michael knew the nosey old Kennedy bitch from the Coll had made at least one phone call.

Jo clung close to Michael.

'Well?' demanded the bishop. He hadn't even said hello to Olive.

Michael said, 'There's not a priest in the diocese who would dare marry us.' He skipped the "your Grace" He felt the Church was about to finesse him with an excommunication.

'I should bloody hope not,' roared the bishop. He turned and took Olive's hands in his. 'I called at the hospital on the way, he's looking much better.' She looked tearful; he sent her scurrying for a cup of tea. 'And some of that apple pie of yours.'

He planked himself down in a chair. 'So you're going to live in sin? Are living in sin.'

Michael put his arm around Jo, her tremble was worst than his, but touching her gave him strength. 'There is a Church law, if we cannot get a priest to marry us within six months then we can marry each other in the sight of God.'

'We already have,' said Jo.

'That's hair splitting,' roared the bishop.

Michael licked dry lips. He said, 'If God didn't split hairs, none of us would make heaven.'

The bishop accepted a cup of tea from Olive with a smile then glowered over at Michael. 'There's a heresy there somewhere if I cared to look for it.'

Michael thought to sit down himself. Immediately the bishop snapped out of his chair and stood over them. 'You're supposed to suffer for God, not put your feet up and feel hard done by when things get tough. I need you here for three months certain, there's nobody else.'

'Sorry,' said Michael.

'Sorry!' roared the bishop. 'Sorry!' He stomped around the kitchen then stopped over them again. 'You can split as many hairs with God as you like, but not with my parishioners. If nobody else will do it, I will. Tomorrow morning, the seven thirty Mass, I marry you. And may the Lord have mercy on my soul when the Vatican finds out.'

He stormed out of the kitchen. Jo started to cry, then Olive. Even Michael had trouble with moisture in his eyes.

The bishop came back. 'Sorry,' he muttered to Olive and went out again. This time the front door banged behind him.

Jo went to Olive and the two women clung to each other. Michael finished pouring the tea for all of them to give himself something to do. He paced about the kitchen, excited at the thought of being both married and a practicing priest. Eventually the women separated and they drank their nearly tepid tea.

Jo said, in despair. 'I have nothing to wear.' She rushed off to tear through her things looking for something suitable.

Olive sat on in the kitchen with Michael. Eventually she said, 'You young people won't want me around. I'd be in the way.'

'It's your house,' said Michael, knowing it wasn't.

She shook her head. 'If Fr Toner stays here he'll keep on working.' An idea occurred to her, she looked pleased. 'Maybe we could go to Norway, I'd love to see it again.'

He asked, 'You've been there before then?'

'With a very special friend.'

Michael recognised the glow of love in her aging face, and had the sense to sit silent while she hunted a packet of photographs out of a drawer. They were all of Norwegian scenery, and all of them featured Olive, as she was forty years before: a young woman with jet-black hair, facing a perfect future. Some of her poses were risqué for their day, and he formed the impression of a very physical relationship. The photographs were of a professional standard.

'They're very good,' he said.

She said, 'He was an artist at heart. All the paintings in the house, he did those.'

She turned up the next photograph and hesitated before she passed it over. 'He wanted me to pose nude for this one, but I couldn't, not outside, not with people passing.'

The photograph showed her dressed in a white slip that drifting moisture had made practically transparent. From the firmness of her nipples and the way the slip clung it was obvious that she was naked underneath.

Michael cleared his throat and asked, trying to appear natural. 'Have you one of your young man?'

She hunted through the pack. 'Somebody took this one of us.'

It was of the same scene, taken further away. The photograph showed the young couple clinging together; they looked ecstatic. In the background was the full waterfall. Olive wore a white mantilla. Michael looked at Olive's lover and thought he recognised the face. It came to him. The young man was the bishop.

THE PRIDE OF HENRY GEORGE

Henry George Osborne-Bradley liked to spend his lunch hour at the Safari Park. He used to meet Susan there until… Well anyway, he still went.

This day, the last Wednesday in September, was cold with a hint of rain in the air. There was no man at the main gate, no "white hunter" to take his money. He began to wonder if the man was ill, lying collapsed and dying in his hut. Heart attack? Diabetic coma? Sunstroke? Brain tumour? Snake bite? Dehydration? The list of ills was endless once Henry's mother delved into her medical books. She'd had the symptoms of them all at one time or another, even the snakebite. Though that turned out to be a stray piece of wood with a sharp edge.

Henry went across and checked the hut. No body lay there to upset him, no broken glass with a trail of blood leading from it. However, a selection of coin sat, stacked neatly, on the table, and a stray twenty-pound note flapped gently in the breeze. He pursed his lips. 'Oh dear.'

Thinking of money and safety made him do an automatic check of his waistcoat pocket. It held Susan's present to him from last Christmas. A signet ring, the curving lines of his initials hinting at her desire for their names to entwine. Just as their limbs had done for a few glorious minutes until his mobile rang.

It rang now. He answered it. 'Yes, Mother?'

'Henry George, Have you been to Courtneys yet?'

'Not yet.'

Her indignation vibrated over the ether. 'I don't like you going anywhere near that place. In future, get somebody else to call. You are the Sales Director after all.'

'Mother, I've told you. What Mr Courtney wants, Mr Courtney gets.'

Mr Courtney was irascible, demanding and long past retirement age. He insisted in dealing with only the top salesperson in any company, the man or woman with the authority to give him the last possible penny in discount. He was Browns' largest customer so he was listened to. Mr Courtney also fancied himself as a matchmaker. Henry George Osborne-Bradley, Esquire, bachelor and youngest director of Browns, to his equally available and long widowed secretary. Too long according to Mr Courtney.

Henry disconnected. He made his muscles relax against the angry burn in his stomach, and started to worry again about the missing man. He wondered if he had been taken short. Irritable Bowel Syndrome, colitis, diverticulitis, Crohn's disease? He looked across to the thick belt of bushes separating the road from the main car park, and called out, diffidently. 'Excuse me.'

There was no reply.

Worried now for the well being of the missing man he hurried across to the bushes. 'Hallo?'

Again he got no reply. And not only that. Other than the metallic hiss of traffic on the far motorway and the rattle of drying leaves in the September air, there was no sound. No bird song, no voices. Not even from the monkeys at the teahouse who always had something to chatter about.

Henry thought of reversing the car back onto the main road and leaving. Then he thought of the man lying hurt or ill in the shrubbery, and pushed his way through, head down, looking all the time for a pair of boots or a hand reaching out from below a bush. He found sweet wrappers and condoms, and a two euro piece, which he pocketed against his next trip to Dublin. Two euros, hardly the price of a cup of tea. He hoped some child hadn't lost it.

Like the child he could hear now. At least he thought it a child, keening deep in its throat. The only other noise was the swish of that year's soft growth on his clothes. Even the breeze had stilled.

He came to the end of the shrubbery and stopped at the kerb bordering the car park. The teahouse, and the monkey house, and the animal enclosures lay at the far end. There was hardly a car left in the car park, and not a person in sight other than the child. A little boy, maybe two years old, standing quiet in his short trousers and a Liverpool shirt. His mouth was haloed with strawberry ice cream. The cone lay forgotten at his feet.

What made the keening noise was a large male lion pacing circles round the little boy.

The lion looked Henry's way, and the keen became a *whuff*. 'Keep clear. Mine.'

'No he is not,' said Henry.

The lion stopped pacing and *whuffed* again. Henry found himself walking, dream state, towards the lion's next meal.

'You're all *whuff*,' he scolded. 'And you don't want that child. You've had your breakfast.'

He kept walking, swallowed his heart down out of his throat. To him the lion looked leggy and a bit thin, an adolescent Leo. It needed filling out. He wished he hadn't thought that.

The lion kept *whuffing*. Henry reached the little boy, who clung to the leg of his light-coloured suit. The little hands were sticky-dirty, and he knew the suit would need dry-cleaned. The lion's uncertain step forward reminded Henry of a vacillating customer. One in particular, the manager of a major Belfast store, who could never steel himself to buy enough from Browns in case he bought too much, and the directors were annoyed.

The thought of that customer gave Henry the strength to look the lion straight in the eye. 'I can see you now. Out in the big world you are an important person, the decision maker.' He snapped his fingers. 'Click, click, click, and everybody jumps to your tune. But when it comes something personal, something nearer home? You're nothing.'

The lion stopped with his paw in the air. 'What's that?' asked Henry. 'Your idea of putting your foot down? You're pathetic. Leo the Pussy Foot.'

He thought he could hear his own bones crackle, and realised something was moving through the bushes. Leo heard the noise as well and sidled away from it. Unfortunately, each backward step brought him closer to Henry. And if Leo was frightened of the noise…?

Henry scooped the little boy into his arms just as a low-slung lioness broke cover. 'What's your name?' he asked.

'Robert. I want to go home. I want my mammy.'

A warm dampness invaded Henry's arm. The little boy had wet himself. Or worse. There was a certain *frisson* in the air. Across the car park he could see a blur of faces watching from the restaurant window. Nobody came to help.

The lioness roared. Leo roared back.

'Mummy!' screamed Robert, and hugged Henry until he thought he would choke. The pad, pad of the lioness' feet on the tarmac came nearly as quickly as his heartbeat. 'Things were bad enough without you,' he told the lioness.

Leo's next roar came weak, feigning anger at the lioness' interference. The lioness had a limp.

'Dot and carry one,' said Henry, and tried to put a tone of confidence in his voice.

He remembered hearing from somewhere that it was the lioness who filled the larder, not the lion. First they manoeuvred the weakling out of the herd, then they cut it off from the rest before moving in for the kill. Dot appeared to be going straight to stage three. He supposed he could always drop little Robert and walk away. He remembered the crunch and slurp as his mother's miniature poodle, Ferdinand de Rocamadour, stripped the remains of the Sunday leg of lamb down to raw bone, and knew he couldn't do that. Anyway, the child held him in a death grip. The stickiness from the little hands had transferred to Henry's neck and made him feel uncomfortable. He had always wanted a child. Nobody had ever said how unhygienic they were.

He stood on. He couldn't think of anything else to do as Dot reached Leo and tried to shoulder him out of her way. Leo snarled at her and gave another roar. He took the lead as the pair advanced. He even licked his lips.

'Oh yes,' said Henry. 'You're a big boy now with a bit of backup.'

Leo looked like Mr Courtney when he was making his mind up to demand some impossible discount. Even if he were wrong, pride wouldn't let the old man back down. The trick was to break his chain of thought before the decision was made.

Henry actually felt himself become angry. He took a step towards the advancing Leo. 'You weren't so hot on your own.' He thought if he lived he would get himself checked out at a psychiatric hospital.

Leo stopped and looked behind for guidance from Dot.

'Ha!' said Henry and took another step. He wondered what he thought he was doing, then he remembered the television programme he had seen years before about George Adamson of *Born Free* fame. Adamson had controlled a pride of lions by shaking a stick at them. A really large pride, two dozen or more, who survived by hunting game in the African bush.

All he was up against was two bottle-fed wimps, who had never attacked anything more dangerous than a plastic bag blown across their compound. 'You think you're bad,' he shouted at them. 'Some of my customers could take blood from a stone.' Both lions were stopped now. Their next roar was almost a yawn; disinterested they would have him believe. He got a hand free from supporting Robert and waved it dismissively. 'On you go about your business.'

And they obeyed him, heads hangdog, passing him in an elliptical curve. He took a deep breath; it went out again like the first breath after a cold swim. His knees wanted to buckle.

'Go on,' he said, in an encouraging, gentle voice, because they were heading towards the lion enclosure.

Dot went with a snarl still in her throat. Susan was the same when he called at Courtneys. Until he took the signet ring she had given him out of his pocket, and wore it – mother's approval or not, mother's palpitations and threatened heart failure not withstanding – then, and only then, would she walk out with him again.

Meanwhile he walked behind the lions herding them on. He felt he had to. If he tried to make a break for safety, distance might lend them confidence. His vision fixed on the huge outer-gate leading into the lion enclosure. It was open, and didn't seem to be getting any closer.

He wondered if more lions might be roaming free.

This thought nipped at the edges of his confidence. The lions sensed it and became edgy, their heads came up and Dot's snarl deepened. They passed a group of cars. People sat in them. Adults frozen, children with their noses squashed against the windows. Somebody was using a camcorder.

Henry tried to sidle in the direction of the cars and safety. At once Dot whipped round, snarling, her eyes burning holes in him. He forgot about the cars, forgot about stuffing little Robert through a window and into safety. Instead, he kept walking at Dot, talking quietly. 'Go on out of here. Mind your noise.'

He was almost a hand-stroke away from Dot when she gave a final snarl, turned away and trotted on to catch up with Leo, who was heading for the enclosure and home.

Henry followed, wondering if the lions were actually leading him to their larder. Eat the boy now, keep him for later. Or was it leopards which did that? Store game in a handy tree and call it home. He thought Dot might know if he asked, but hesitated. He didn't want to be giving her any ideas.

He tried to get a grip of himself. He knew his nerve was starting to slip and he was verging on hysteria.

'Mister, you're sweating,' said Robert's little voice.

The lions broke step at the tender, soft-boned sound. Henry's heart nearly stopped. 'Shush,' he said, quickly.

'But you are,' insisted Robert.

Henry was so hot his sweat ran like basting liquid. Add browning and serve up, he thought. He hated meat cooked rare, but even more he detested the way his mother served up vegetables boiled tasteless. Susan, he discovered, cooked them with a crunch. That was after he had finally built up the nerve to tell mother about Susan, and taken up one of her invitations to dinner.

Her dessert he hadn't even tasted. 'Just a nibble,' he said, too full of main course to be greedy. But somehow his "just a nibble" took on another connotation. Perhaps it was the twitch of her eyebrow; perhaps it was the morning-pink flush rising on her cheeks. He knew she was wondering the same thing as him. How the evening would end. Their hands met, their lips. They made the settee and their clothes came loose, were pulled away. Then the phone rang.

Not mother this time, the neighbour. Mother had been rushed to hospital. The evening wilted, he had to go, rushing to dress with embarrassed apologies.

The family doctor was a man well versed in mother's ways. His diagnosis agreed with Susan's. 'A severe case of galloping histrionics.' He shook his head. 'How dare you leave your mother for an evening's tête-à-tête with another woman?' He gave Henry a friendly thump on the chest. 'You'll be talking marriage next.'

Dot kept looking back, eyeing him. Leo walked on, his glances to the side fearful. Henry felt proud that he had the young lion cowed. Then came the awful realisation. Leo wasn't looking at him, but beyond.

He sneaked a look in the same direction. Coming in from the side was a second lioness. Something dangled from her jaws. She stopped and tried to spit it out. It stuck. She shook her head violently and spat again. The body of a miniature poodle splattered onto the ground. Its little pompom tail drooped with gore.

It made him mad the useless destruction of the little dog. 'You're not even hungry,' he told the mother lioness. 'Or you'd have eaten it in two bites and a snap.'

The mother lioness was nearly as big as a donkey, and solid as a carthorse. The way she came at him reminded him of his mother. At least the lioness intended to kill him clean. Mother had destroyed his spirit and left the husk of his hopes to serve her.

He said, 'You'll find no spine, it was snapped out of me a lifetime ago.'

He touched the bulge in his pocket that was Susan's ring. He hated the thought of the mother lioness consuming that as well. The ultimate victory for the old…. He stopped himself. Henry never cursed. Not because mother disapproved, but because his father had hugged him gently and said, 'Son, I'd rather you didn't use words like that.'

Henry hugged little Robert, and willed himself not to back off from the lioness in his last moments. He wished he could find some way of saving the little boy, even as he found his feet taking reluctant steps forward. He didn't remember his brain giving the instructions, but it had, and it kept on giving his feet the order to advance. The mother lioness glared at Henry the way his own did, but without the same hate in her eyes. Killing was an instinct to her, something she did. It was nothing personal.

He stood that bit straighter. This wasn't going to be too bad, a moment's pain. Perhaps not even that. He wished he could have said goodbye to Susan. Susan whom mother disapproved off, 'Not nearly good enough.' Henry often thought he would gladly swap Susan's hot genes for the Osbornes' cold logic.

Dot came circling behind him, ready to join in the kill. She pawed at Robert. Henry swung round and tried to slap her snout. Robert screwed his head out from under Henry's chin and saw Dot backing off, spitting angrily. His little trembles became deep shudders. He was going to cry out loud. And if he did? 'There, there,' said Henry hastily, and patted the little back. The tremors deepened. 'Liverpool are rubbish,' he added. Robert stiffened with indignation, and the cry died in his throat.

In all the manoeuvring Henry found himself hard up against Leo. Leo seemed glad of his company.

'Stay the way you are, son,' said Henry. He had this crazy idea of putting Robert on Leo's back and letting the lion carry the little boy to safety.

He kept walking towards the mother lioness, who stopped and watched him out of narrowed eyes, not sure of what she was up against. She gave a roar, a deep, meaty roar. Leo's reply was almost a whimper and Dot sounded angry at the world in general.

Henry kept walking. He stopped a breath away from the mother lioness and stabbed a finger at the lion enclosure. 'Get to hell in there before I put my toe up your arse.'

He had never ever used words like that before in his life. It felt good. Just the right thing to say, the right tone.

The lioness glared at him, he glared back. It came to him. Like George Adamson, he was no longer afraid, and fear controls most actions. Robert sensed the change as well. He giggled, a bit nervous, but he giggled.

Henry couldn't bring himself to say it, not in front of the little boy, so he moderated the first word. 'Feck off out of my sight.'

The mother lioness turned and walked away. Dot followed. Leo stayed with Henry who wished him gone. Little bubbles of growls built up in Leo's throat until he sounded quite savage. 'Get you,' said Henry and tickled him behind the ears.

Leo went all stiff at first, then he rested his head against Henry's leg. It was like getting hit by a truck. He staggered and nearly fell.

The mother lioness sensed an advantage and whipped round.

'Feck off,' screamed Henry, fear coursing through him like a pain.

And she did. Through the outer-gate, and into the mesh-wire

tunnel leading to the lion enclosure. Dot trotted after her, snarling at everything, then a reluctant Leo. He was going to get hell when he got home.

A mysterious hand pressed a button and the electrically operated gate slid shut. It was over. The monkeys started to chatter among themselves

An hour later Henry walked purposefully into Courtneys. He ignored Mr Courtney at his desk and stood over Susan.

'Mother's in hospital,' he announced.

'Oh.' She didn't sound unduly concerned.

'The doctor thought it best. Palpitations, hysteria and heart failure.'

'Oh?' she said, again.

'Yes. I told her I was probably going to wear your ring.'

Fury burned in Susan's cheeks. 'Only "probably." Nine months and it's still only "probably."'

Henry slapped the signet ring down on the desk between them. He said, 'It depends whether you want me to wear it on my finger, or through my nose?'

WAITING TO GO ON

Mummy went away because she loves you. And you mustn't smile, it's sad, Mummy going away. Sam says she won't be back, but we'll see her tonight. I hope she'll be waiting for us and not be busy or something. I don't like the smell, do you? And I should have put your dirty nappy out first. I left it for Sam; he likes playing mummies and daddies. He says I worry too much, but everything is such a muddle.

It was so much nicer when it was just Mummy and me. I don't mean you, you're lovely, I wouldn't swap you for all the world. I love your little burps, you always smile afterwards, and I love the way your lips uncurl and stretch nearly to your ear. It's a pity they don't work on the other side, and I'm sure both your eyes would have been brown. That one's got the most gorgeous yellow dots in it, much better than Mummy's or mine, and Sam has got none at all. Isn't that funny? Miss George, she's my biology teacher, she says it's got something to do with sweet pea. She explained it in class the other day, how the bees going from flower to flower mix up the colours; the way you do when you try to eat the daisies.

Miss George doesn't shout at me when I don't understand things. I had lots of homework for tomorrow. I've got it all done, even the French verbs: *j'aime; tu as*, though they won't see them because they're in my head. If you're hungry there's a bottle left, you could have it while we go, but I don't want you to be sick over your good dress. It doesn't matter about the blanket that was mine, and it's been washed dozens of times.

I should be dancing tonight. Miss Hubbard will be cross when I don't come. You know Miss Hubbard; she gave you the green duck. She was much nicer when Uncle Don was around. She sucked up to him all the time, and she went to the Chinese at Christmas because Uncle Don told her we were going. It's funny that, Mummy made me call him uncle and he's not, but Sam's Sam and he's our grandfather.

Anyway, I had Coke, and Uncle Don and Mummy and Miss Hubbard drank out of a big black bottle, the whole lot, right down to the bottom. Then they had another one. Maybe that's what made Mummy sick? Miss Hubbard got quite giggly and Uncle Don had to run her home. Mummy went straight to bed with a hot-water bottle; the pain was so bad she cried. I know, I heard her through the wall.

I don't like Uncle Don, but that's a secret. You're not to say. It's because he cheated. He said I wasn't growing, but I am, my dresses are getting shorter. Mummy met Uncle Don in a pub. He came to live with us after Sam and Mummy had a fight and Sam left. It was terrible. Mummy screamed at Sam because he wasn't wearing any trousers, and threw me out of the room. You should have seen the bruises on my arm, big black dots they were. Sam said that Mummy was jealous, and banged the door when he left.

I ran to my bedroom and cried as hard as I could. Mummy came and said she was sorry, she didn't mean to hurt me; and she kissed the bruises better. I tried to join them up with paint while she phoned the police, but it ran. After that Mummy cried and cried. She told me that men were bad and I was never ever to speak to them when I grow up. She made me promise and I did, boys are so *yucky* anyway.

The police came, and looked under the beds and in the wardrobes. I told them Sam wasn't there, that he got stuck under the bed once when we were playing hide-and-go-seek. The policemen laughed and let me play with their radios while they drank tea. I talked into one, and a lady at the other end answered. The lady had a funny voice, but she was nice. She promised to have a good look for Sam and she found him right away. Wasn't that clever?

Sam didn't dare come home because Mummy was still cross. The police found his pictures on top of the wardrobe; not even Mummy knew they were there. They had a good look and tut-tutted. They said there was one just like me, but Mummy shook her head at me and I said no.

I wanted them to find Teddy when he was lost, but Mummy said they were too busy looking for other granddaddies. We found Teddy in the bin. His leg was all torn and his insides were falling out; Mummy sewed it on again.

Poor Teddy, he was all dirty and wet with tea bags and I was afraid Sam would be too, wet that is. Mummy told me not to worry, and Uncle Don said he was in 'Do not pass go.' Then Mummy and Uncle Don had the biggest row, all whispered shouts, in the kitchen. I listened at the door, but I couldn't make out the words. I wanted Uncle Don to leave as well, but he didn't.

Mummy's often cross; it's her back she says. She gets it from the way she stands. Honestly you'd think she was trying to touch her toes with her nose. I told her that one day. I pretended I was Miss Hubbard; I stood the way she does when she tells us things. Miss Hubbard stands very nicely, but she limps when she walks. Some days she's very limpy and very cross.

Well anyway, I told Mummy, I made my voice all squeaky, just like Miss Hubbard's. 'Poise and deportment, girls. Poise and deportment.' Mummy tried, but she kept forgetting. If it was me, Miss Hubbard would make me do *point and toe* exercises round the room for weeks and weeks, and not let me dance ever again until my whole body could stretch to the stars. And she wouldn't let me dance with the big girls. I know I'm twelve... well nearly twelve, I will be in ten days, and Carla my friend isn't dancing and she's nearly thirteen. She says, 'Who cares anyway?' And she's not coming to see me dance. She says she's too busy.

Sam says I'm the star of the show. When I run onto the stage everybody stops looking at the big girls. They all say, 'Oh', and I get the biggest clap. I curtsy to the seats in the front row where all the important people sit. I think Sam should be there as well, he comes every night and they only come once.

I do the prettiest curtsy. I make sure my hands catch my tutu so that I look like a swan sitting on its eggs, like I saw in the park one day with Uncle Don and Mummy. That was the day Uncle Don took my height and said he made two of me. He didn't have to cheat the first time, stand on his toes I mean.

Then after the Oh and the clap, Sibyl takes me by the hand... I must have told you about Sibyl. She's been dancing for years and years and she never falls off her toes, not once. Sibyl's going to Saddlers Wells. She wants me to come. She says I can sleep in her room and carry her bags between classes.

Well anyway, Sibyl leads me to the centre of the stage. I curl up as small as small. She gives me a pole with ribbons to hold while the big girls dance round me. Their feet get closer and closer, and I think they're going to step on me the way Uncle Don steps on people he doesn't like. He says they never get up again.

I never let on I'm scared. I close my eyes tight until the clapping starts then I give everybody a big smile and run off. Miss Hubbard doesn't say who the smile should be for. Every night I look up and give it to Mummy. I know Sam wouldn't mind, he never minds anything nice about Mummy.

Uncle Don calls him Sad Sam. I heard him say it to his new girlfriend one day in the street, and she laughed. I don't think that's nice even though Sam went away and missed my birthday, and sometimes Mummy cried so much she had to take medicine, and then she couldn't wake up to make my tea. Sometimes she fell down.

Sam's never sad when we cuddle together, and he only hurts a little. He always smiles afterwards and calls me his little princess. I asked him to help me with my French verbs. He says he only knows *Brigit Bardot*; it was a joke once, about the time his mummy and daddy died. He says he can't remember them. I think that's so sad, not remembering. I'm glad we won't forget Mummy. Uncle Don was always saying how people should do something. Why don't they give children like Sam new mummies and daddies, instead of putting them in a big house with cross men to look after them?

Sam is never cross with me. Uncle Don shouts a lot, especially if you don't do what he says. He shouted the day he moved in with us. Mummy said no when he said he was coming to stay, and he got so mad. Honestly it was *awful*. I hid under my bed until he poked at me with the brush to come and get my tea.

Sometimes Uncle Don got me ready for bed. I felt a bit funny about it and made him promise not to look while I changed my you-know-whats. But he did, honestly, he did, I saw his eyes hiding behind his fingers. I told him he was naughty, and he blew bubbles on my tummy until I promised not to tell. I was really glad when he pushed me down the stairs. Mummy hit him with a saucepan. She told him to leave and he banged the door as well. I'm not allowed to do that. Sam came back to live with us, and ages after that you arrived.

It's a year today since Mummy went away and it's your birthday.

Miss George remembered and she had a special present for you. I left it behind. Sorry. I'd been thinking about the sweet peas, they didn't seem to have anything to do with people, and I asked Miss George about it. Miss George was very nice. She made me sit down, and she cried as well. She wanted to take me to the doctor, but I sneaked away.

Mummy, and you, and me. Sam has visited all our flowers. And my baby? It might have no eyes at all. I shouldn't be talking, you're asleep and I will be soon. You look lovely; I never saw pink in your cheeks before. It must be the gas.

THE FLY POOL

He hesitated on the doorstep, looking worried, the man who had raped her. He asked. 'Will you be all right?'

'Yes,' said Cora. Calmly.

'I mean....' He edged towards his car. Stopped with his hand on the door handle, and tried to hold her gaze. 'I'll ring.'

Cora ignored his need for reassurance. She went back into the house and closed the door, used fingers to work her throat where it hurt from crying. She wore no engagement ring; the wedding band was dull.

Out front the car started up. She heard it crunch slowly down the gravel drive, wait at the gate for traffic to pass. Pull away.

The house looked... different. This wasn't her home, not with the arc of grime under the hall table and the stink of cigarette smoke in the air. Corners and under things were Mrs Millar's domain, and they hadn't been touched in a long time.

Cora found herself in the kitchen, and went from cupboard to fridge to freezer, looking not touching. Checked the tops of things for grime, and flicked it off her fingertips. Left a note for Mrs Millar.

> *Gone to do the shopping. Hall, sitting room, kitchen.*
> *Give them a good clean. Pull things out.*

She knew she should mention the smoking; the children's rooms stank of it. She had even found Leona with a butt in her mouth. While she hesitated, the front door slapped open and Mrs Millar banged into the hall. Lladro shuddered on its base and the chandelier tinkled. The sharp aroma of fresh cigarette smoke invaded the kitchen.

'Mrs Millar, not in the house please.'

Mrs Millar stopped in the kitchen doorway, her hands full of empty shopping bags. The cigarette stuck out aggressively. 'You never said before.'

'I did. Please don't do it again.'

Cora found herself trembling. Mrs Millar was a dream cleaner, tackling dirt the way she tackled authority, but only after you motivated the anger that had bad-mouthed her out of every other job. Cora decided, with relief, that she had said enough for now. She picked up an envelope of money from the table and tried to make her escape.

Mrs Millar blocked her way. 'That's the shopping money.'

'I'm going shopping.'

'I do the shopping here,' insisted Mrs Millar.

Cora wanted to play Tarzan, the way she did with the children. 'You clean. House dirty,' and knew she was panicking. The tremble under her breastbone started up. Her throat hurt. 'Not today.'

'I need things. Cigarettes, our Willie's tea.'

Cora held out her hand. 'Give me the money and I'll get them for you.'

Mrs Millar hesitated, her face tightened hard. 'I forgot my purse.'

'Then how…?' Cora was conscious of the envelope of money in her hand and decided not to finish that question. She hurried out of the house without a coat.

Luckily it was warm outside, an extended Indian summer after the children had gone back to school. What breeze there was had a nip to it. Good fishing weather, she thought, though all she had ever done was throw food-pellets into the water to encourage the fish to rise. It never seemed to work, but it was fun watching the ripples run out from the splash.

Cora pulled her cardigan tight round her and walked quickly. Up the road she remembered the car in the garage. She hadn't had it out in weeks, months, April, in fact. There was ice, she remembered, the car sliding sideways on the drive. She had run back into the house in panic. After that, a taxi took the children to school and brought them back in the afternoon.

She felt out of breath and slowed her pace, slid fingers under her cardigan and moulded her stomach muscles, shocked at their softness. Something else she didn't do now, walk miles to kill the loneliness, and go to aerobics.

She noticed fresh paint on two of the shops, the red of the newsagents dusty already from passing cars. She nodded to people she knew from meeting them in the street. They nodded back, their foreheads creasing as they tried to remember when they had last met her and who she was. Past the bank, the post office, the office supply shop, now empty – when had that happened? - with a sun-faded notice on the door giving a forwarding address. The building society on the corner with another broken window. 'Some things don't change,' said Cora to herself, and sighed as she turned down the entry to the supermarket.

The supermarket was quiet at that time of the morning. More people sat at tills, or walked the isles with clipboards, than did any actual buying. She recognised one of the staff mopping the floor, though he had grown inches and filled out at the chest. She didn't have to see his staff badge to know his name.

'Jonathan, good morning.'

The youth turned round. He said, 'Good morning, Mrs Vallely.' He held the mop handle up, defensive. Looked like he wanted to run.

Cora felt her own face burn as she remembered the last rush of shopping before Christmas. It was the badge, with her son's name, Jonathan, on it, coming at her unexpectedly as she turned the corner into the household section. She had cried then, banging her head off the terrazzo floor to take away the pain. She touched the youth's arm. 'Jonathan, it's a nice name. "Beloved of God."'

Taking pity on his embarrassment, she walked over to the vegetables: potatoes, green, red and yellow peppers. Chives. Broccoli. Cora's mouth watered, she hadn't tasted broccoli since Mrs Millar had taken over the shopping, and fresh vegetables were restricted to carrots, parsnips and onions.

She had forgotten a trolley. She went back and got one, and dithered and delayed round the displays. Lamb steaks, she thought. With roast Mediterranean vegetables and couscous. The price of things now shocked her. Even with the envelope bulging money, she wondered if she had enough. She had, with plenty to spare, and thought to buy Mrs Millar her cigarettes. Finished shopping, she sat in the adjoining café and drank a coffee. The longer she stayed out the better. It let Mrs Millar take her venom out on the house, leave things scraped pristine.

Then he came, the rapist, slipping into the seat beside her. She looked to get away, but he had her trapped against the wall. She decided not to make a fuss in public.

He asked, 'What are you doing here?' He looked strained.

'Shopping for fresh food. I'm sick of Mrs Millar's tins and frozens.'

'Oh.' He relaxed a bit. 'You're not going…?'

'To the police? No.'

'Good.' He wiped sweat off his face, he had been hurrying, and sipped at her coffee, which had cooled. Made a face at its bitter taste. He liked his well-sweetened.

She said, 'You had an important meeting this morning.'

'That's what I pay staff for.'

'And you were late in?'

'Well after last night….' He stopped.

'And now this?' It surprised her, him making excuses for not working. She took the cup from him. 'You could always get your own coffee.'

'I suppose.' He sat on. He was afraid she would run.

She looked him straight in the eye, said it deliberately. 'You raped me.'

He looked away. Tore open a sugar sachet, spilled it. 'I didn't mean to. You started to scream when I was coming… I couldn't stop myself.'

She interrupted. 'You wanted me to have another Jonathan.'

He went all red in the face, close to crying. 'Oh God! I'm so sorry.'

Her lips tightened until they went white. 'You were never there for him, "Beloved of God."'

'He was Down's.' It came out as a pain.

'Your son.'

'Yes. Yes.'

'So you gave me a big house and a fancy car, a chequebook and no questions asked. All I had to do was *cope*.' The final word she spat at him.

She wanted out, her own air to breathe. She pushed at him.

He said, 'I'll take you home.'

'No.' Her legs were tired, cramped, like she had been sitting forever at the fly pool, letting her imagination ride the ripples to the far bank. 'Yes.'

She walked on, leaving him to deal with the bags. Outside in the car park she looked for the Mercedes 500s, then remembered it had gone. The Audi estate was much more child-friendly and he didn't mind the mess so much. He swung the back lid up and put the shopping in among the children's swimming gear. No fishing rods, no boxes of lures and flies, to take up the space. He looked thinner, had a different tiredness under his eyes, and the once careful coiffeur was now the cut-and-go hairstyle of a local barber. He towered over her, immensely strong, yet when he helped her into the car it was done with a great gentleness.

They set off, out of the car park and down the road, working their way through a one-way system, which didn't always make sense. He tried to make conversation. She replied to each question with "yes" or "no" still angry with him for following her. The foot-well had a layer of sweetie papers. *He* didn't allow the children sweets.

She pursed her lips and glowered at him from under her eyebrows, the way he did when crossed. 'It's no wonder they don't eat their tea.'

He gave her a side-look and bit down a smile. Swung to the left at the junction.

'This is the long way,' she said.

'I thought….'

She said, 'It's been a year. More.' Sat back in her seat and closed her eyes, remembering Jonathan. The big trusting eyes, the ready smile. His last words. 'Look, Mummy. Lorry.'

Her eyes snapped open. The police station was coming up. 'Pull in,' she said.

He kept driving.

She insisted. 'Jim, you've got to trust me.'

He swallowed hard, choked down his feelings. 'Cora. Please.'

She undid her seat belt and put a hand on the door handle.

He braked and swung to the kerb, panicking. 'All right, all right. Don't do anything stupid.'

They pulled up right outside the police station. There was a policeman standing at the door. Cora recognised him though she didn't know his name. Number 3741. He had brought her a cup of tea on one of her tearful visits. She demanding to be prosecuted. Jailed.

Jim was tearing his way out of the seatbelt and the car. The policeman was on the radio. 'It's that bloody woman, Vallely, she's back again.'

Cora put a hand on Jim's arm. 'Trust me,' she said, and held on until he pushed himself back in his seat. He sat tight as a double-snapped stick. Cora got out of the car and nodded to the policeman, who touched a finger to his cap. Her legs seemed to have melted down to her knees. She found it difficult to walk steadily across the wide pavement, past the police station and into the flower shop next door.

She had intended to buy only one flower, a lily, but the pots and the colours drew her and she filled her arms with fragrance. On the way back to the car, with Jim rushing to open doors, she stopped at the policeman. She said, 'The accident was my fault, but you gave me sympathy instead of punishment. Thank you.'

He touched his cap again, not knowing what to say.

Jim came to help, breathing hard as if he had been running. This time he held the door for her and let her get herself in. 'Trust,' he said. A sigh of relief puffed his cheeks. 'Did you ever hear of heart failure?'

He saw the white lily held reverently in her lap. Without asking, he turned the Audi and headed back towards the cemetery, where they stood together, not touching, at a grave marked with a dove of peace.

Below the hill of the cemetery and across the field beyond, was the river where they had first met. He casting into the fly pool from the depth of the evening shadows. She strolling along the path until a midge got in her eye, and he rushed to help. Seeing the river was almost another pain to her. They needed to get back to those early days again, him fishing. She dibbling fingers in the water, dreaming of their perfect life together, until he shouted. 'Did you see that one jump?' She never did, all she ever saw was the ripples.

Cora allowed herself one tear for Jonathan, but refused to cry for Jim's sake. For the pain and fear lurking in the eyes of this man, her husband, who had tried to force a child on her to help lift her depression. She studied him, as if meeting for the first time, and saw the father who drove a family-friendly car, who went swimming with their children in the evenings, and took time off from pressing work to be with her. The man who never had time for hobbies, for himself.

She said, 'Last night - afterwards - I dreamt of Jonathan. He was so happy in heaven I cried because he didn't miss me.' She looked over at her husband. 'There were times I thought you didn't need me either.'

He held out a hesitant hand. 'I'm here now.'

She covered his hand with hers, locked fingers. She said, 'James, "Son of Thunder." You always were. I just didn't know it.'

GROWING PAINS

People say, 'That's her, that's Znell,' and some ask, 'Are you all right now?' but their eyes stay carefully away from my feet... foot. Now Vince, he finds me in the hospital trying out my new foot. He knocks the crutches away and waltzes me up the corridor.

Our feet tangle and we fall against the wall. I chase after him and try to kick the lining out of him. Then I stop. I'm on my own with no crutches? I sink to the ground and start to cry.

Vince comes close and bends over me. He says, 'You've got natural balance, poise. Use it!'

I scratch his eyes out. He puts a tissue to his face. A nurse sees the blood and takes him to the treatment room. He asks for a rabies shot.

That's Vince, always gets the last word. You've no idea how it grates. So what am I doing outside the Summer Palace Theatre with the rain beating off the estuary? I'm standing looking at a poster, that's what.

STARING
VINCE BUCHANAN

It's not much of a theatre and the town is hardly better. Always before I wore a short belt of fancy material over skinny knickers, and high kicked my fans into a later. Tonight I'm reduced to bell-bottom trousers because my stump aches in the cold.

Vince and I first met at TV talent auditions and, later, during rehearsals for the finals night of the series. I was barely eighteen and he was much older, twenty-four or something. I won, of course. My agent, one of the big name, high-power type, was already talking about Vegas and Carnegie Hall. Vince was a little talent going a long way. I had to slap him down a few times over the week.

'What are you going to do with the prize money?' somebody asked me. And loudmouth butted in. 'Make an opening wide enough to get her head out the door.'

The theatre door opens and Vince stands there. 'Susan!' he cries, all dramatic, and gives my mother the biggest hug. She's as bad back. Those two can chat for hours. Even a cup of tea in the canteen takes them forever.

All the older women are like that with him. And I don't know what he gets up to in the children's ward. Once, when the screams and shouts got too much, Sister marched in and came out with her uniform creased and her perm everywhere. 'Who can't do a handstand?' she said, and walked off with her face screwed tight against a smile.

I haven't done Carnegie Hall, but I did make Vegas and the charts. I met Vince again at a party, and pretended to vaguely remember him. 'Oh Vic! Like the nasal spray, always up your nose.'

He used that line on television the very next week, and he made number nineteen in the charts because he bought everybody a copy of his record for Christmas. Mine made number one. His wasn't bad actually, it rather grows on you, and his second hung around in the charts for weeks. Singing underwater for Children in Need and nearly drowning must have helped.

Anyway, we met again at this party. I was trying to have a sensible conversation with a Frenchman, who matched his professional desire for *The Voice*, as he called me, with designs on my body. French endearments are not quite so romantic when they have to be shouted over the sound of a noisy Conga.

Vince was in the lead of course, twisting and turning the line through the room. He managed to plough into the Frenchman just as the man's hand slipped below belt level. My hopes for a French tour ended up on the floor. I didn't mind really, a fortnight spent fighting off *Monsieur l'Impresario's* hands is not my idea of fun. I've no time for that sort of thing anyway, but I wasn't going to let Vince off with it.

'Are you ever serious?' I asked him.

'It's my dour Scotch blood, I have to keep it in check,' he said.

'Scots. Scotch is a drink,' I told him.

'Not if you're a vampire,' he said.

My mother was there that night. He took her for a waltz while the band played heavy metal music. She came back all flushed and her cheeks glowing. 'He does it all the time,' she said.

'Does what?' I snapped, furious with her for being so embarrassing. After all, I have an image to maintain.

'Ask me outside for a smooch.' She fluffed her hair up at the back. 'Someday I'll call his bluff.'

And I had brought her along to keep preying hands off me?

Eventually Vince and mother finish this great play-sexy greeting. I hold out my hand. 'Good evening, Vincent.'

'Good evening, Violet Elizabeth,' he says and lead the way backstage.

I hate my given names, he knows that, but he doesn't know I'm so nervous I'll want the toilet in a minute. This is my first public appearance since the accident. I could have gone to any show in the West End, taken a bow, and been swamped in champagne, tears and sympathy from my peers. Sympathy is the last thing I'll get here tonight.

A lot of performers are hanging around, most of them kids, wide-eyed and trying not to look under my bell-bottoms. I want to run.

Vince says, 'You must have lost weight, your trousers aren't arse tight.'

I grit my teeth and give the kids a nod of the head. Vince leads us to a settee in the wings. It's all chintz and flowers, and has been around longer than Methuselah. 'My office,' he says, and clears a clutter of paper onto the floor. The kids follow us and hang around, but none come too close.

'They're very disciplined,' says mother.

'They're under threat of death if they annoy,' he says, and goes off to organise things. Vince Buchanan, star, compére and bullyboy. I know I have made a mistake. Wanting to see if I, Znell, have the drive to take me back to the top is one thing. Starting rock bottom again is something else.

Mother sits down and makes herself comfortable. 'I always wanted to try a casting couch,' she sighs, and flutters her eyelashes at Vince. He blows a kiss back.

Sometimes I wonder why I bother bringing her along.

I look out the side of the curtain and see the seats full of grannies and mothers, and a host of teenagers. Far too many for a wet night on the sands. People see the curtain move. They point and the shout goes up. 'Znell! Znell! Znell!' Nobody is supposed to know I'm here.

Vince is back. I say, 'I charge five thousand for an appearance.'

He says, 'A walk-on and bow doesn't even qualify for Equity rates.' He wanders onto the stage, the kids trip after him and the show starts. I think I will leave at the interval.

It's a good show by well-established second-raters, and children from the local dancing schools for the *ahhh* factor. Not that I have any interest, but you can't help noticing things. Like Vince walking about backstage in between introducing the acts. He is apparently quite casual, but behind him everything goes click, click, click into place.

When he's not doing that he is standing in the wings, singing with the singers, gyrating with the dancers, virtually doing handstands with the tumblers. There is always a kid holding his hand. I think he is more entertainment than the acts. I think he would make a great father.

The kids get closer to me; nudging each other and saying 'Go on, I dare you.' Vince comes to shoo them away and mother tells him to leave them alone. A little blonde-haired minx mimics her, 'Go 'way, 'ince.' He does. I don't mind the kids and their wide-eyed stares. I started the same way.

The interval comes and I stay on. I'm in pain, but I hide it. I have this muscle behind my knee. It jumps then twists itself into a spasm of agony. It's an old injury, all I have to do is flex my foot and wriggle my toes and the pain is gone. But so is my foot, and I talk to the kids to distract my mind.

Vince brings mother a cup of tea. He has to lean over the kids to give it to her, and puts a hand on my leg for balance. His fingers slide into the damaged muscle and the pain goes. I don't know how he does that, know when I'm in pain and cure it with a touch.

The first time it happened was in the hospital. According to the nerve-ends in my leg, the foot was still there and hanging over a blazing fire. Then the muscle started to jump. I was lying, sweating out the physiotherapist's lunch break, when Vince walked in. He talked for a while then his hand went under the covers.

That was all I needed, a groper taking pleasure in handling my raw stump. I told him so in a few choice words. Besides I was only in a tee shirt and panties, and I hadn't showered that morning. I rang for the nurse, but by then his thumb was working miracles on the muscle.

'How did you know?' I asked him.

'It happened to you finals night of the talent competition, you wouldn't let me help then.'

I remembered telling him to bugger off and mind his own bloody business.

The release of pain was like an orgasm; my burning foot stopped hurting as well. I felt myself sinking into a deep sleep. He kissed my lips. 'Sleep well, Princess,' and left.

Vince follows up mother's tea with a glass of peach flavoured water for me. Still water to keep down the gas when dancing. The second half starts, and the cast work miracles on an audience wanting to be pleased. Vince is running now, shooing people here and there, helping the band set up their instruments at the back of the stage. Then rushing off to change for his big number.

He is good, his routine less aggressive and more confidant. I am trying to place one of the songs; it is a variation of one I know. Then he rips off his trousers and dances around in hot pants. My dancing is always spontaneous, and a devil for the cameras to follow. Not my problem. His is choreographed, and what he does with his tongue and hips you would only see in Soho. The audience loves it. I think they are going to throw their knickers.

Then I recognise the song. It's my first big hit. He is mocking me.

Vince comes off the stage. 'I hate you,' I say, but he doesn't hear me over the thunder of applause.

Mother frowns at me. 'Why do you dislike Vince so much?'

'I despise him because… because….'

'Because he treats you as a human being and not an idol,' she snaps, and gives Vince a big hug and tells him how wonderful he was.

I cut out the hug and the adulation and say he was good, and, yeah, he had the audience in the palm of his hand. Just listen to the calls for an encore. He looks pleased and I don't mention the mockery bit. Why give him an opportunity to gloat?

The curtain comes down and the comic is doing a quick turn out front while the cast prepares for the grand finale. And for my walk on and bow. Vince suggests I have a dry run, and gives me his microphone for balance as I walk out onto the stage. Walk! Run, dance, explode perhaps, but Znell never walked onto a stage in her life.

I feel strange, out of kilter, and my steps get slower as I near my spot, the white X marking centre stage for the star. I close my eyes and remember the good old days, the rush of ecstasy as I prepared to sing, my now missing foot itching to kick high. I find myself humming my first hit, the real version. I hug my arms around me as the music comes, a slither on the drums, a vibrated G-string.

I turn and look behind me. The band is in their seats, picking their instruments. Tears come to my eyes and my memory downloads the words to my lips, the quiet start, the raunchy, hip-lusting climax. I begin to sing, softly, caressing the words. The backing group take up the hum. I hold the microphone to me, all but eat it with suppressed lust, turn and....

The curtain is up. It's not supposed to be. And the audience are watching, motionless, like they have been frozen in time. I am going to *kill* Vince. I don't have to sing, damn it! I can walk off stage.

The music keeps going, so does my voice, and my feet are tapping the rhythm. My singing isn't good, not with the dust of the hospital still in my throat. I miss the top B and hesitate over the lyrics, but already I am better than Vince. Internationally better.

In between verses I do a bit of fancy footwork, forget myself, try a high-kick and fall. Getting to my knees will be a humiliating crawl. Vince comes high kicking across the stage. He looks ridiculous dancing around me in white tie, tails and pink hot pants, and his face is slick with sweat. He reaches down, grips my fingertips in his and floats me to my feet. He dances back off stage.

The band holds the beat, wanting me to try again, and I do. I need weights in the boot to get the real height, and I'm afraid my foot will fly off and hit somebody in the audience. I start the second verse.

The audience is screaming, they couldn't possibly hear a word. They are leaving their seats and coming to the front of the stage. I don't do grannies, but they know the words and sing along. The cast comes onto the stage as I start a second song. The older troopers keep the kids out from under my feet. Vince stays in the wings.

The second song is hard and fast, and high-kicked me into super stardom. I'm not good and I'm going to be sore tomorrow. I think my stump is bleeding. I don't care if the foot flies off. I fall twice more and bounce to my feet almost before I can hit the stage. The song finishes. I collapse in a heap centre stage, hold my arms out and gather the kids to me. They smell of innocence and I envy them. The curtain goes up and down a million times.

The kids help me to my feet and I limp into the wings. They think I'm playacting and laugh. Vince is there, still sweating. He turns away from me and I realise that the sweat starts below his eyes. That he is crying. I go to him. He hugs me. 'Princess,' he croaks. 'Princess.'

His hug is filling space and needs in me I never knew existed. I think Vince, like his song, is growing on me.

DOUGHNUTS

Kenny cut out the opposition. This fare was somebody special. The woman jumped back as the taxi slewed to a halt beside her. She recognised Kenny, and tried to flag down the other taxi instead, but it drove on.

Kenny got out to help her. 'Long time no see,' he said.

She began to look pleased, then embarrassed, then pleased again. She busied herself with the child while he put the buggy in the boot.

'Where to?' he asked, when he was back at the wheel. He checked her and the road in the rear-view mirror. She was sitting on the edge of the seat, still fussing over the child. She looked older; the round face he remembered had lengthened into maturity.

'Castle Court.' She kept her head down. 'I'm meeting my husband there.'

It was easy to keep the smile on his face; she always had that way with him.

He said, 'I haven't seen you in a while. Things going well?'

'Yes, yes.'

He watched her in the rear-view mirror as much as possible. There was less of the strict Evangelical about the way she dressed. Wool and worsted had given way to silks and suede, and the tight hair bun, cropped and soothed to an alluring sheen. 'Same old taxi,' he said. 'Still a good ride.'

Colour burned in her face, she wouldn't look at him. He laughed.

Her head came up, the mouth tight with temper. 'Kenny Smith you're no better.'

'And why would I change?'

'Why indeed?' She put the baby on her hip and sat back in the corner.

Suddenly he found the smile difficult to hold. He hadn't looked at the child - boy or girl, he didn't know which - but it could have been his. He looked at the mother again and wouldn't have minded.

There was the silence of shared memories between them until he pulled up at Castle Court. He fetched the buggy and made room for it on the pavement among all the people, and held it steady while the child was being clipped into its harness. It was a girl, with blue eyes like her mother's, big and trusting, and blonde hair going dark. About eighteen months old, he thought, though babies weren't his scene.

'Do I know your husband?' he asked, to break the silence between them.

'No, I met him in England.'

His eyes were drawn again to the child. 'And this is the perfect baby, reared by the instruction manuals.'

'What would you know about it?'

He pointed to a sign at the door. 'I bet you've never given the kid a piece of doughnut.'

'Fast food…'

'I know, I know.'

Her fingers trembled as she hunted her purse for the money. She said, 'You're at it again.'

'And you're Miss Perfect.'

He noticed something; he stilled her hands and put his thumb and forefinger where her wedding ring should have been. 'Has this paragon of virtue you married got a name?

'James.' Her face burned. 'James.' It came out stronger the second time.

Kenny didn't believe her, there was no ridge on the ring finger. He looked at the child and tried to work out dates and times; sweat broke out over his body.

She slipped her hand from his grasp. 'Goodbye,' she said firmly.

He blocked the buggy with his foot. 'You never said?' He felt he was making the biggest fool of himself.

'You were having too much fun.' There was anger in her voice.

He had to admit to himself that she was right, but felt he had changed since they last met. Casual sex no longer interested him. Most of the women he had grown up with were married or in relationships, and he had nothing in common with the rest.

He hunkered down beside the baby. 'Hello, Little Jo.'

'Josephine! And how did you know her name?'

'It's yours, isn't it?' He brushed the baby's cheek with his hand; it was like touching silk. He recognised his own family's stubborn chin and saw a spark in the young eyes that he liked. Textbooks or no textbooks, that young lady was going to have an opinion of her own. He stood up and took the mother's hand again, and held it as easily as his heart held her memory. He found himself smiling. She smiled back.

They went into the shopping centre and shared a doughnut with their daughter.

THIS LIFE

Jimmy and Doc were brothers. They had worked as a team from the moment they stood side by side in the headmaster's office, wondering why they had been called and what it had to do with the lanky workman who stood in the corner, screwing his cap round and round in his hands.

The headmaster made it as easy for them as possible. Their father was dead.

The *Sick* was no option for a man determined to educate his sons out of the squalor of working class Belfast life. Jimmy senior had ignored his occasional bouts of breathlessness and gone to work. One moment he was in the middle of Mackies foundry, shouting at his mates to heave on a length of steel girder, the next he was lying at their feet.

The headmaster nodded at the workman who stilled his hands. 'Michael Quinn here, brought the word. He also brought your father's last wage packet. They gave you the week even though it's only Wednesday.'

Something boiled up in Jimmy through the pain of his loss. Before he could make sense of his confused feelings Michael Quinn stepped forward and shook his hand. 'I'm sorry for your trouble,' he said.

Jimmy left Everton Secondary School immediately and took a job in Harland and Wolff as a trainee welder. His mother was already failing, and after a few weeks he transferred to the carding room in Ewarts' mill so that he could run home at lunchtime and see to things. The neighbours were very good. They tried to help, but they had their own problems and, eventually, everything but the most intimate attention to his mother fell on his shoulders. Mornings were a rush, the evenings filled with doing housework and helping Doc with his homework.

That winter was wet, and the compact terrace house became chronically damp and impossible to heat. Their mother's asthma flared up. Jimmy tried everything including anger to keep her alive, but she died with the change of year.

The day after the funeral Jimmy took Doc by the hand, and they walked the long miles across the divide and up the Falls Road to the City Cemetery. They stood over a rectangle of mud marked by a simple cross; Jimmy had his arm around Doc's shoulders. Doc muttered something.

'What?' asked Jimmy.

Doc said, 'It's a line from Pasternak: life is not a stroll across a field.'

Jimmy's grip tightened as he swore vengeance on a system that had condemned the two people he loved most to an early grave. But who or what he wasn't sure.

★

It was nineteen seventy and the Civil Rights campaigners were tramping all over Northern Ireland. They said they were ordinary God-fearing people demanding their civil rights; Jimmy's contacts in the police told him different. They said it was IRA inciting the Catholics into another round of violence.

Jimmy had learned a lot in the intervening years. A short stint in Mackies had ended with him taking a swipe at a supervisor. The supervisor was bigger and heavier, and knew how to hit. That was Jimmy's first lesson: keep your hands in your pockets until you have sussed the opposition. After that he laboured wherever he could get work, mostly on building sites along Lower Library Street where they were tearing down old housing to put up flats for the Catholics.

That annoyed him: the waste, the dumping of perfectly good building materials. He started coming into work early and leaving late. He used that time to strip Bangor-blue slates off the roofs and stack them in a corner. A chippie tipped him the wink and he collected roof joists as well. When he thought he had enough to make a load, he dropped a lorry driver a few bob to run it home for him. Only the wood made it, the foreman, Craig, diverted the slate for his own benefit.

'You young bucks think you know it all,' gloated Craig, when Jimmy went storming in the next morning.

Jimmy kept his hands deep in his pockets and well out of trouble. He said, 'I did it in my own time. It was stuff you were dumping anyway.'

Craig made a point of keeping his hands in sight; they were the size of soup plates. 'That's as maybe.'

'Maybe what?' asked Jimmy. He knew a deal in the offing when he saw one.

From then on he spent much of the day retrieving materials for the two of them, him and Craig, to sell on a fifty, fifty basis. He still came in early and left late, and, every so often, he would have a friend bring over a lorry and they'd load it up with the more valuable items, ones Craig never knew about.

Things worth retrieving kept getting broken. 'I never thought,' someone would say.

'You never thought in your life,' retorted Jimmy.

Then a lorry scattered a load of slate waiting to be collected.

The driver said, 'It shouldn't be there anyway.' He just happened to be holding a wrench in his hand at the time.

Jimmy stabbed a finger. 'I'll see you in the pub after.'

'You and which army?' said the driver. He didn't look unduly worried; he could have made two of Jimmy in any direction.

'Salvation Army,' muttered Jimmy and went off to think up a strategy. He had calmed down enough to realise that suicide wasn't a reasonable option.

When he got to the pub all of the men from the site were there to see the fun. Jimmy's nose wrinkled at the sour smell of bodies and stale drink. The driver and two equally large friends advanced on him 'You were saying?'

It was a Friday and payday. Jimmy put his wage packet, unopened, on the counter. 'I was saying, your money's no good.'

'What's wrong with my money?'

'Give me strength,' said Craig, who was there as well. He dumped the driver on a stool and told the barman. 'Give brains here a pint and chaser.'

'Drinks all round,' said Jimmy.

'Good on you, son,' said Craig, and gave Jimmy a big wink. It was all he gave. When the money ran done he excused himself and left.

Every week after that Jimmy put his wages on the bar counter, and from then on there were fewer breakages and things were put aside for him to pick up later.

Evenings and weekends Jimmy traded second hand materials out of his backyard. There was always somebody fixing up a room or building a pigeon loft, and not against a bargain. Jimmy sold the wood, well matured yellow pine, at a third of the price of cut white wood from the sawmills, a quarter if the purchaser was willing to take the nails out himself. The Bangor blues had the same sort of discount and he learned to haggle over the rest.

Doc was still at school. Jimmy had everything mapped out for him. 'Sums, there's always an opening for a good clerk.'

Doc shrugged. He was into the school play and existentialism in a big way. He stayed back most evenings to practice and had a photograph of Sartre on the bedroom wall.

Craig got to hear of Jimmy's shop and came visiting the little terrace house. He walked around, mean in his silence, and did a bit of counting on his fingers. Jimmy made a show of unconcern, and kept his hands jammed into his trouser pockets and out of trouble.

'I'm impressed,' said Craig, eventually.

He was entitled to be. The shop had grown out of the backyard into the kitchen, and from there into the parlour and one of the bedrooms. Even then it was difficult to get around without moving things.

Craig held his hand out. 'My cut, we must have made quite a bit.'

'You think so?' Jimmy took his hands out of his pockets.

Craig bellied forward, pushing him against a cast iron fireplace. 'Fifty fifty.'

Jimmy's heart was pounding. 'That's not the way I see it.'

'Now look, son.'

Jimmy felt along the wall for something heavy, nothing lethal, something that would take out a few teeth. 'I'm not your son. Back off.'

Craig pushed harder. 'I don't want to be unpleasant.'

A cricket ball whistled past his head. It hit the inner front door and the glass shattered. Doc hefted a second ball.

Jimmy said, 'That's my little brother. He plays cricket for the school, and he only misses when he wants to.'

Doc hefted the ball again. Craig backed off. 'It's your last chance. We share or I get myself another partner.'

'I'll pick up my wages Friday.' Jimmy spoke with more bravado than he felt.

'I wouldn't waste the trip.'

Craig left. Jimmy rubbed his back while he buttoned down his despair; good jobs were hard to come by. He looked at Doc. 'You did mean to miss, didn't you?'

'Why?' asked Doc.

On Friday Jimmy gave Doc a day off school and timed his arrival at the site for three thirty, pay time. He made Doc wear his school uniform and picked every piece of fluff off the jacket before he let him out the door.

Craig was expecting them. So were the rest of the crew. They all crushed into the doorway of the site hut to see the fun. The site hut was long and thin, the wooden floor sagged in places and tools lay scattered along walls half eaten by damp. The wages clerk sat as far back in the room as he possibly could. A desk and a box of wage packets were his only defence.

Craig was in front of the table; he stood with his arms folded. 'Jimmy Terence, nothing due.'

Jimmy looked about him from the corner of his eye. He felt like a trapped rat, but could see no way of backing down. 'Yes there is.'

'Fuck off, sonny, before I give you a belt.'

'Lying week and a couple of days.' He gave undue emphasis to the 'Lying.'

Craig hit him, an open-handed slap across the cheek that was more insulting than an honest blow. Jimmy went staggering back. The men held him up so that he didn't fall. One of them hissed, 'Stay on your feet. That bastard has steel-capped boots, and he's quick to use them.'

Jimmy's shook his helpers off. He flexed his shoulders and bunched his fists. He couldn't understand the fool pride that stopped him from running.

Craig advanced, his boots loud on the floor. Jimmy jumped forward, ducked under a swinging fist the size of a cannon ball and hit Craig. It was like hitting concrete, his knuckles started to swell.

'Fight dirty,' shouted Doc.

Jimmy threw a right-hander. Craig blocked the blow and hit him above the heart. The world stopped for Jimmy, the follow up blows was all sensation and no sound; he couldn't hear, he couldn't breathe. Then he was lying on the ground watching a toecap draw back.

'He's only a kid,' shouted the man who had warned him earlier.

Jimmy heard that.

And Craig's reply. 'He'll not make any older.'

'For God's sake!' Doc sounded impatient.

'You try it,' said Jimmy back. At least he thought he did.

Craig went for the groin, Jimmy just managed to roll outwards and take it on the leg. He cried out in pain as his thigh bone jumped in its socket. The boot was drawn back again.

None of the men were willing to risk helping Jimmy. Doc picked up a nail bar and swung. He wasn't very tall then so he went for what he could reach. The first blow shattered a shoulder blade; the next broke an arm and drove a rib into Craig's lung. Craig went down.

'Stop it,' said Jimmy. This time he knew he hadn't spoken. He wished he had left Doc at home.

The other men grabbed Doc as he went for the head. He fought them, silent and white faced, until Jimmy got enough breath together to order him to stop. Doc still held the nail-bar, everyone backed off cautiously.

Jimmy used tools and the wall to get to his feet; his left leg wouldn't lock at the knee. He held out his hand. The clerk rushed to give him his money.

Jimmy gave the packet to one of the men. 'The drinks are on me.'

He limped out. Doc followed Jimmy and the men followed Doc, leaving the clerk to call the ambulance.

★

The Troubles came to Belfast in August 1971. Jimmy was eighteen and doing rightly: bricks and wood, fireplaces and sinks were his capital. If you wanted it he had it, and if he hadn't he got it; either way it came cheap. Doc was still at school. In his spare time he was reading *The Outsider*, 'In its original French,' Jimmy told everybody with all the pride of a doting mother.

Jimmy's eyes popped open. Something had wakened him. It came again, the wail of banshees from hell. 'Bloody TV,' he murmured into the night. He knew he should go down and play the heavy brother, but he was warm and comfortable and didn't want to move. He pulled the blankets over his head and went back to sleep.

He heard it again. This time he came storming awake. 'I'll kill him.'

Doc was already in the room. 'Did you hear that?'

'Would you go to your bed?'

'No, listen.'

They heard more cries; a distant light flickered against the drawn blind. Jimmy kicked himself clear of the blankets and snapped the blind up; the cord spun itself around the roller. He didn't notice because he was looking at a house burning two or three streets away. He grabbed his clothes. 'Somebody must be trapped.'

'Right, said Doc, and disappeared into his own room.

They met on the stairs, Jimmy slightly in the lead, and raced out of the house. Their neighbours were already in the street or crowding their doorways.

Jimmy met Sam Gilliland; he was swinging a crowbar in his massive hands. 'What's up?'

Sam said, 'The IRA are coming.'

In the distance a sharp crack sounded. Sam grabbed Jimmy and hustled him against a wall. 'That's a 303.'

'Three o' three, what?'

'Rifle you idiot.'

'Fuck.'

Something banged into the roof above them, a piece of slate slithered into the gutter.

Sam's grip was tight on Jimmy's arm. 'Get your stuff and clear out, this is no place for honest men.'

'It's my home, I'll see them in hell first.'

Sam pushed him towards his house. 'And they're the men to send you there. Fall back and regroup, then we can advance and retake the position.'

Jimmy tried to make a laugh of it. 'You make it sound like war?'

'Welcome to the real world son.'

Sam went off to do his own packing. Jimmy hesitated. The glow from the burning house was showing over the roofs, and women, who normally wouldn't be seen with even a hair-grip missing, were at their doors with only a blanket held round them for decency sake. The street was crowded with people yet the greatest noise was that of the wind rushing to the flames and the crackle of burning wood.

A woman came flying over. She was carrying a bundle of things in a blanket and held two children firmly by one hand. 'Jimmy, Jimmy, Jimmy.'

'Yes missis?'

'Could you give the wanes and me a lift to Sandy Row? I've got a sister there.'

Jimmy hesitated. All he had was a flatbed Bedford truck, hardly bigger than a motorcar. It would take him several runs to get his own stuff clear: the few bits and pieces left by his mother and some job lots of china he hoped to sell for a good price. He looked down at the children, they were tear-stained and snot nosed.

He knew what his mother would have said so he did it for her. 'Get on board, missis, I'll be with you in a minute.'

He ran into the house, Doc followed.

The woman's voice followed them. 'Jimmy Terence, you're the spitting image of your mother, a decenter woman never lived.'

Jimmy shoved Doc to the stairs. 'Pack and be quick about it.' He shouted after him. 'And keep the lights off.'

He himself risked a flash of light in each room. From the living room he took a picture of his parents on their wedding day, and another of himself and Doc in a silver-gilt frame. There was nothing he wanted in either the kitchen or scullery. In the bedroom he scooped all the clothes he could find onto his bed, threw in the pictures and tied everything as securely in the candlewick bedspread. Next he shoved the bed back against the wardrobe, searched under a loose floorboard and pulled out a thin wad of notes, white fivers, and a little bag that clinked. It held gold sovereigns. He stuffed them into his pocket and raced downstairs again. Doc was waiting for him in the hall with the picture of Sartre under his arm.

'Is that the best you could do?'

'School's out,' said Doc.

The woman and her children were already on the lorry. Jimmy looked at another half dozen who were pushing and pulling each other on board. Jimmy asked. 'Where do you lot think you're going?'

'I'm on your way,' said a wee woman with a sharp nose and even sharper tongue. She hadn't said a civil word to anybody in years.

'More like, in our way,' muttered Doc.

'God bless you, Jimmy,' said somebody among the crowd.

Over the rooftops they saw another house gout flames and sparks soar into the air. The women in the truck crouched low and held their children to them. Another family got on.

Jimmy stuffed Doc and his picture into the passenger seat and jammed him in solid with the bundle of clothes. Doc was grinning. 'You've got a warped sense of humour,' snapped Jimmy.

'And you only get mad when you're worried.'

'Shut up.'

The lorry started, it didn't always. It rattled and banged at the best of times, shuddered like a wild beast when it went off a cylinder, and created its own smog of burnt oil wherever it went. That night it ran like an angel.

The men cheered them on their way; the noise from the back of the lorry was like the middle of a good wake. They turned the corner and the wailing increased.

Two streets on the lorry started to lurch over potholes in the road where the cobblestones had been lifted. The people they passed, mostly men, had fixed stares and no greeting. Many held broken cobblestones in their hand, others axe handles; in one or two cases the axe head was still attached. Beyond that the streetlights were out. The smell of smoke hung heavy in the air.

The lorry's lights picked out a line of men blocking the road. Jimmy pulled up smoothly so as not to startle them.

One of them opened the driver's door. 'Out.'

Jimmy said, 'We're Protestants. I'm taking these women to Sandy Row.'

'We need your lorry for the barricade.'

Jimmy looked over their heads and saw a line of cars lying on their sides across the street. He might refer to the lorry as an "old bitch" but she meant a lot to him. Sometimes he wished his mother was alive. On a good day he could have taken her into the hills to breathe pure air. Maybe even the seaside, Bangor wasn't that far away. 'Over my dead body.'

'That can be arranged,' snarled one of the men, and they closed in on him. Axe handles were hefted; somebody opened Doc's door and poked at him through the bundle of clothes.

The women came pouring off the lorry.

'You're not taking it.'

'The man's running me and the weans to Sandy Row.'

'You expect me to walk and me eight months gone?'

The men backed off, not a retreat exactly, just a step or two that allowed Jimmy to reverse out of the street.

Once turned, he stopped and waited for the women to re-board. They got on muttering:

'Flipping cheek.'

'Who does that lot think they are?'

'We soon sorted them.'

Jimmy said to Doc. 'You were a great help.'

Doc got a hand free. The blade of a kitchen knife gleamed dully.

Jimmy was angry. 'This is no time to get stroppy.'

They drove on. The side streets were no longer safe, Jimmy turned onto the Shankill Road proper and roared down the hill. They were stopped at a police checkpoint near the junction with Peter's Hill, this time it was Lee-Enfields and Sterling sub-machine guns. Torchlight glared in Jimmy's eyes, another washed over the women on the back. They were waved on.

Jimmy leaned out the window and pointed at the Sterlings. 'You couldn't lend me one?'

'Get your own,' said the sergeant. He was only half-joking.

'I might just.' Jimmy thought he should talk to Sam Gilliland, he had the oddest contacts.

They drove down Royal Avenue. The place was neon-lit and eerie. Only the odd police car passed, there were no pedestrians. At Castle Street junction, the bottom of the Catholic Falls Road, the lights were against him. He stopped. A car came up behind him and ran the lights.

'What the hell,' said Jimmy, and went on.

A line of men guarded the entrance into Sandy Row, some held Union Jacks nailed to sticks, others were more threatening.

Jimmy stopped well short of them and shouted at the women. 'Off quick.'

The men advanced, the women took their time. Jimmy sweated, Doc started to wheeze.

The line of men stopped. The man who had given them the order to halt, kept walking; he carried a flag on a stick. Jimmy started to whistle Colonel Bogie but, there was something about the man that was more dangerous than funny, and the notes died on his lips. There wasn't much difference in their respective heights because it was a small lorry and the stranger was a very tall man. He was broad and fleshy, clean-shaven in spite of the late hour, and his skin was so smooth it glistened in the near dark.

He said, 'You're Jimmy Terence.' His accent had more than a hint of the Home Counties in it.

'Am I?' said Jimmy.

'You are. You've done a bit of work for me.'

'No I haven't. I don't know you from Adam.'

'You have you know, you just didn't know it.'

'Well I hope you paid me.' The women were off at long last. For all their hurry to get to Sandy Row they didn't seem to know what to do once they got there. Jimmy jerked his thumb in their direction. 'What about them?'

'We've the church halls open and the kettle on. As many as you can bring we can cope with.'

Jimmy thought of the silent men lining the streets and the barricades going up. 'There's not much chance of that.'

'Tell them the Man said it was all right.'

Jimmy started to laugh, felt the steely eyes boring into him and cut it off. He tried to cover. 'It might take a bit more than that.'

The Man turned and indicated, it was the most regal wave Jimmy had ever seen outside of the Queen. Two men came forward. One of them was carrying a second flag; the other had a limping slouch of a walk. It was Jimmy's old adversary, the foreman, Craig. Craig looked everywhere but at Jimmy. The Man took the flag off the first man and handed it in. 'Tie it to a wing mirror.'

Jimmy passed it over to Doc, who muttered, 'What do you think this is, Crackerjack?'

A lot of the loneliness Jimmy felt in the night disappeared, he was more respectful now. 'You couldn't back that up with a Sterling sub-machine gun just in case it's the other side?'

The Man spoke to Craig who held out something else for Jimmy. It was a revolver. 'Not him, the other one,' said the Man.

Craig took the gun round and it handed in. The butt slipped neatly into Doc's hand, it almost clicked into place.

On the way back home they detoured to Jimmy's 'shop', the stables of an old brewery, now defunct.

The gates swung open as they approached, a man poked his head out. The man was stubby built and not particularly bright. His name was Peter Dawes; everybody called him Peter Paws because he had one obsession, dogs, Doberman Pinchers. Two adult dogs stood close to his feet and he cradled three pups in his arms as he spoke. 'I thought it was you, the old bitch is running well tonight.'

'Don't even mention it,' said Jimmy, and drove on into the yard. Doc got out, Jimmy sat on. Beyond the lights of the lorry was darkness as far as the distant high-rise flats; one of them was on fire. The houses surrounding the yard were bricked up and derelict. Jimmy loved the place; his impossible dream was of owning it some day.

Doc had disappeared. Jimmy followed the sound of a sledgehammer on concrete and found him knocking in the rear window of the largest house. He was making hard work of it.

Jimmy settled himself against the return wall. 'What do you think you're doing?'

Doc paused briefly. 'Opening the door to our new home.'

'Aye,' said Jimmy. 'Aye.' He couldn't bear to watch a handless worker doing what came easy to him. He took the sledge off Doc. 'This way.' Three blows and the window was in, five more and blocks, glass, wooden frame, the lot were lying in the kitchen floor. He stepped through the window and struck a match. The room was musty and there was the sour smell of mice. It was still furnished: a table and a few chairs. The match went out, Jimmy struck another one. There was an old newspaper in the hearth, he put the match to that and light flared.

Doc was waiting with the bundles of clothes. 'Aye,' said Jimmy, again and took them off him. Doc came back with twigs, Jimmy threw then on the fire, and by the time they were burning Doc was back again with a half bucket of coal. 'You'd make a great wee housewife,' said Jimmy.

They collected tarpaulin and rope and went on.

★

The street, when they got back, was half blocked with bundles and furniture. Jimmy had to snake the lorry in and out of the stuff, and people were forever wandering in his way. The smoke was thicker, the fires closer and his neighbours fears were now edged with panic. Some of them tried to load furniture and family onto the lorry before Jimmy got it stopped.

Sam came running with the crowbar. He cracked it down on the lorry bed. 'None of that nonsense, there's room for all.'

Jimmy nudged Doc. 'Bring the beds and the bedding. And some clothes this time, school uniforms don't grow on trees.'

Doc flashed a smile and went off, he was wheezing badly in the smoke.

Jimmy turned to find himself facing the most wonderful pair of dark eyes.

She was older than him by nearly two years, and when you're eighteen that's a long time. The dress was to the ankles, the cardigan to the knees, they were colourless in the poor lighting, and the white blouse was stapled to her neck by a cameo broach. She was holding a plate of sandwiches.

'God bless you.' He went to take one.

The plate was pulled back. 'We are going to wash our hands first, aren't we?'

He looked at her stupid. His body reeked of sweat and effort, and his hands were black. He could have died of embarrassment. 'You're quite right, missis, though yours look clean enough to me.'

'Pardon?'

'You did say "we."'

'Really?' She flounced off and offered the plate to somebody else, she didn't ask them to wash first, then she looked back at him and smiled. For that smile he would do anything. He went off and washed his hands, and his arms, and his face, and his neck. He found a fresh shirt in the kitchen press and put that on.

He couldn't find her when he went back into the crowd. He fixed a smile on his face and helped Doc and Sam load their things onto the roof of the lorry. He felt angry with her for being a tease. Somebody shoved a mug of tea into his hand and he gulped it down.

He turned and found her standing right behind him, the plate well depleted. Her voice was tart. 'You did remember under the nails?'

His smile became real. 'And ears.' He took a sandwich.

'My own,' she told him.

'Delicious.' He had no idea what he was eating. He was cringing, his shirt was un-ironed, he never bothered with such niceties, and he hoped she hadn't gone into the house to see how they lived.

He waved the mug in her face. 'I know you. You're Eleanor somebody from up the Avenue where all the posh accents come from. All front gardens and Venetian blinds.'

'And you're Young Jimmy Terence. At least your father had some manners. He knew to say thank you.'

He coloured, he choked. She offered the plate again and he took a second sandwich. He struggled to thank her through the choking. Her teeth gleamed as she laughed at him.

He caught his breath. 'You're awful tall for a woman.' She was on the pavement; he was standing in the gutter.

She said, 'There's not much to you. Five nine?'

'And a half.' The half was important to him.

She leaned forward. His hands were tied up with a mug and sandwich. He ached to hold her.

She said, 'Good kissing height.'

Their lips touched. He thought he would die of happiness.

The last trip was near dawn. They were tiring and the lorry had little petrol left.

Jimmy's eyes were burning with the effort of keeping them open. 'Talk to me,' he said to Doc.

Doc was sitting upright and alert. Only his eyes moved, searching through the yellow pall of smoke that hung over the city. He seemed to be enjoying himself. He said, 'I think everybody's gone.'

'Could be.'

'And the shooting's stopped.'

'They're tired, God help them.'

They were two turnings off home. Jimmy took the first and peered down a street he had known from childhood, and was unsure if he was right. It was burning, house upon house, and a body lay in the gutters. He was shocked. The lorry jerked to a halt.

'What are you stopping for?' said Doc.

Jimmy pointed.

'Drive on,' said Doc.

'You're a cold bastard.'

'He's dead, it's dangerous. Drive on.'

They glared at each other. Jimmy said, 'You don't know that for sure.' He had his hand on the door handle.

'His head's half stove-in, take a look.'

Jimmy looked; with the moving darkness around the flames he couldn't be sure. 'I'll check anyway.' The sudden blast of heat from the flames as the door opened made him cringe back.

Doc leaned across and shoved the gun into his hands. 'You'll need this.'

'This is Belfast, head-bin, not the wild west.'

He stuffed the gun into his pocket and went on, it saved arguing.

The heat from the flames was intense. Jimmy had to hold his arms across his face as he approached the body, and his trousers were uncomfortably hot against his legs. The man was dead and was cooking nicely. Jimmy backed off rapidly, what was left of the sandwiches heaved in his stomach.

Something smashed into his back and upper arm. He stumbled towards the flames. Another blow landed before he could turn to face his attacker. This one scraped along the side of his ear and caught the edge of his shoulder; his left arm flopped useless.

There were three of them, and each held a curved stick with a large flat head. Shillelagh? Hurley? Jimmy wasn't sure which.

One of the men gave him a hard poke in the guts. Definitely a hurley, Jimmy thought as he doubled up in pain and went back a step.

'Cook the bastard,' said one of the men.

The hurleys jabbed again. Jimmy's foot turned on a cobble and he almost fell. He could feel the hair on the back of his head curl in the heat and his sweater was smoking. The men were laughing.

He remembered the gun and pulled it out of his pocket. They didn't see it or wouldn't be frightened. He fired. The middleman flopped onto his back; his hurley somersaulted through the air.

A hand, Doc's, came out of the dark and caught it. The other two men were standing stupid. Doc stepped back and swung.

Jimmy closed his eyes, but couldn't blot out the sound of bone crunching.

★

Doc was standing over Jimmy. 'I thought you were going to see Eleanor?'

Jimmy got an eye open. 'I am, I had my bath.' The words came out slurred.

'You need an early night. Or maybe that's what you had in mind?'

'Shut your face.'

The world started to clear around Jimmy. The kitchen with the old tiled floor and a couple of comfortable chairs came into focus. They had electricity, and gas, and a back door to come and go through. Everything was as clean as they could make it because he was thinking of inviting Eleanor round.

Doc shoved somebody forward. 'Whippet's here.'

Whippet made Jimmy feel tall. He spoke high-pitched and quick, it came out like a yap. 'The Man wants you.'

'Does he?' said Jimmy, thinking that it wasn't his night; Eleanor wouldn't come anyway, she wasn't that sort of girl.

'You and your lorry, something's come up.'

'I'm busy.'

'Best advice, come at once.'

Jimmy came fully awake. 'Is that a threat?'

Whippet backed off rapidly. 'I'm only saying what I was told.' He sidled into the scullery and slipped off into the night.

Jimmy was raging with frustration. He had not heard from the Man in a year, not directly, and now this summons, and on a Friday night too. He slammed his feet into his boots, grabbed his coat and went out. Doc followed.

The old lorry was in a bitchy mood. 'I thought you could drive this thing?' snapped Jimmy, after Doc stalled her for the third time.

Doc looked amused, Jimmy got more uptight as they neared Eleanor's house.

Eleanor wasn't too pleased, especially when he could only talk vaguely about business and the Man. She stood in her hallway with a full-length dressing gown held tight to her throat. It was a nice dressing gown, all roses and flowers; Jimmy wished she would open it wide. He left as soon as he politely could; the chill was getting to him.

Doc quoted:

Her voice was ever soft, gentle and low, an...

Jimmy pinned him to the door by the throat. 'Listen, smart ass, if I kill you let it be for something you made up yourself.'

The Man was equally aloof. There had been talk.

'Not me,' said Jimmy.

The stare he got matched Eleanor's for coldness. 'If it had, you wouldn't be here... You wouldn't be anywhere.'

Doc stood silent at Jimmy's shoulder. His hand slipped to the knife he carried. It was made of fine steel with a tip like the point of a needle.

If the Man saw the movement he ignored it. 'As I was saying, when I was so rudely interrupted, there's been too much loose talk. Certain items have to be moved. Tomorrow will be too late.'

A pulse beat Jimmy's head. 'When you say certain items?'

'Yes.'

'I haven't asked the question yet.'

'Don't.'

They stared at each other. Jimmy felt he would always be looking up at the Man, even from the top of a stepladder. He compromised. 'If I don't know?'

'It won't save you.'

'Fuck.'

Jimmy wasn't sure that he wanted to get involved, not that way. He was glad to break stare with the Man, and look back at Doc as if asking his opinion. There was a gleam in Doc's eye.

It was the same gleam Jimmy had seen as they stood over the dead bodies in the burning street. They had never discussed that night, let alone told anyone else, though the Man suspected something. He had checked the gun when Jimmy returned it and saw that a bullet was missing.

'Yes,' said Doc.

The smell of the burning buildings, and the fear of the women and children still haunted Jimmy. He nodded and felt chill.

'Right,' said the Man, and gave his orders quickly. He had a map of the city already spread out on the table. The finger pointed. Jimmy was to collect here, deliver there. Then...

Jimmy swallowed hard. 'How many runs?'

'Three should do it.'

'I'm not a Kamikaze, you know.'

'And I'm not a chicken, I don't put all my eggs in the one basket.'

Jimmy pointed to the jail on the Crumlin Road. 'When you're sending the cake with a file in it, I like marzipan.'

'Icing, too,' said the Man, and led them to an old stone warehouse near the Oranmore Bridge. The foreman, Craig, was waiting for them; his face was as sour as the twist in his shoulder.

Doc reached for his knife. Jimmy stopped. 'Wait a minute.'

'Have you and Craig a problem?' asked the Man.

'Yes.'

'Not tonight you haven't.'

Doc's hand was on his knife. Jimmy glared at Craig. 'If anything goes wrong there'll be more than me not around.'

Other men waited in the near-darkness, cigarettes cupped in their hands. They were all family men because this was no night for young Turks with loose mouths. Jimmy hung around and checked for weight distribution while nondescript packages and boxes were loaded onto the lorry; he tried not to touch them. There was a lot and yet...

'Hardly enough to go to war with,' he said.

The Man's nose pinched at the implied criticism. 'Nobody likes us.'

'Aye,' said Jimmy. 'It's easy being a revolutionary, everybody can empathise with that.'

The Man put a friendly hand on his shoulder. 'Empathise? Empathise? Where do you get all these big words?'

Jimmy shrugged, casual. 'Doc has all these books. Sometimes I forget to buy the paper and you've have to have to read something while you eat your tea.'

The lorry was loaded at last. The Man said, 'Good luck.' He didn't ask insulting questions like: if things went wrong, did Jimmy know to keep his mouth shut?

The lorry was running like an angel. 'Don't forget the marzipan,' said Jimmy.

'See you in hell,' said Craig.

The first run was short. The number of army and police Landrovers loitering in the area made Jimmy nervous, but they weren't stopped.

Back at the warehouse the Man and Craig had disappeared. 'They said they'd see you around,' said the new loading boss.

Jimmy paced nervously while the second load was being hoisted aboard. The night might be quiet, but darkness hung over the city like a pall and he felt cold. He was glad to be off. Their destination this time lay across everything that was dangerous, into a wasteland of crumbling houses on the Crumlin Road. They detoured by the city centre, it seemed safer somehow, and there was less chance of them running into a checkpoint. Jimmy's heart was in his mouth every inch of the way.

Doc looked at the police Landrovers stacked around Carlisle Circus. 'They're heading into the Shankill.'

'Don't be daft, it's the Taigs they're after.'

They turned into the Crumlin Road.

Doc said, 'So far...'

'Don't even think it.'

Something caught Jimmy's eye as they passed, a movement in a side street. 'Did you?'

'Yeah.'

He hesitated, his foot eased on the accelerator and the engine stalled.

They came out of the lorry running. 'This is crazy,' said Jimmy as they turned the corner into the side street.

They were right; they had seen the dark shadows of people against the poor street lighting, two on the ground and a third standing over them. Jimmy shouted and two of them, men, ran off, one of them at a hobble as he pulled his trousers up. Jimmy held an arm out to stop Doc from following.

The third person lay on. A woman's voice was gasping, 'Jesus was lost and Jesus was found. Jesus was lost and Jesus was found.'

By the time Jimmy reached her she had managed to push her dress down as far as her knees. White underwear, and a shoe, and other bits of clothing lay nearby. Her hair was dark on the pavement and was as long as Eleanor's.

She groaned.

'Are you all right, love?' said Jimmy. He wanted her to be. The last thing he needed was to get involved.

He struck a match, her hair was rust red in the uncertain light and her face was dark with blood.

'Fuck,' he said. He added, hastily, 'Sorry missis.'

She was a nun. He had never spoken to a nun before, he wasn't sure if they were really human. This one was beautiful.

'Cover her legs, you're not supposed to see them,' said Doc.

'You do it.'

'I'm not snogging Eleanor.'

'Shut up.' Jimmy's hand hovered uncertainly over the material.

'Don't you dare,' said the nun. She burst into tears.

They looked at each other, helpless. Jimmy was sweating. If anything happened to the lorry, if the police came...?

'Missis, I don't want to rush you but we've got our own problems.'

The sobbing stopped. 'Go away.' It started again.

'Oh f...' said Jimmy.

They gave her a minute then Jimmy risked taking an arm and encouraging her to her feet.

Doc said, 'A couple of cars just passed the bottom of the street.'

Jimmy used the word internally and gave the nun a handkerchief to wipe her face. The handkerchief was clean; he had taken it out of the drawer especially for his date.

'I think they've stopped,' said Doc.

The nun shook herself free of his hand. 'Go away.' She meant it this time.

Jimmy understood. Eleanor liked privacy to settle her clothes after their goodnight snog. Usually he had to wait outside the lorry cab, often in the rain. He wondered how many years she would be willing to wait for him.

'We'll wait at the corner.' He doubted if they'd get much further.

They walked slowly; Doc dumped his knife down a grating. When they reached the corner they found men all round the lorry. Jimmy said, 'No matter what you say, say nothing.' Doc was silent, Jimmy felt him tremble. 'You'll be all right. Blame me for everything.'

Torches swung in their direction, they blinked and squinted in its sharp glare.

'Do you own this lorry?'

'I do, officer.'

'It's a bit late to be out with a load.'

'Are you telling me?'

Doc nudged him hard. It wasn't the police, it was the IRA, and they had guns.

'What's your name?'

'Paddy John Joe,' said Jimmy. He was sure he knew that voice from somewhere.

Strange men crowded them, objects, some hard others prickly, were pushed against their bodies and torches glared in their eyes. Hands searched their bodies. They found the driving licence in Jimmy's hip pocket and weren't particularly gentle around the crotch. Jimmy choked back a cough.

The voice said, 'Jimmy Terence, a good Prod name if I ever heard one. And from the Shankill.'

One of the other men growled, 'We haven't got time for this.'

'Now now, boys, we talk to them first then we shoot them.'

Jimmy bit his lip and wished he was somewhere else, wished he had found time to make it with a woman, any woman. Eleanor. A gun barrel was pushed in under his cheekbone. He wanted Doc to make a run for it. He knew the stupid bugger wouldn't go, not without him.

The beam of a torch played along the lorry and focused on a long box. 'What's in there?'

'Sten guns.'

The beam next settled on to a squat box. 'And that?'

Jimmy was sweating hard. 'Dynamite.'

A gun barrel stabbed his guts. 'Smart-arse Prod.' Air hissed out of Jimmy.

The voice said, 'Take a look, Sean.' The lorry springs squealed and the dark shadow of a squat man stood out against the sky.

A knife was pressed hard against Jimmy's throat. 'What are you bastards up to?'

'Helping me,' said a voice. Jimmy jumped, he'd forgotten about the nun. It was a relief when the lights swung in her direction.

The nun stood at the corner. She was nearly as tall as Jimmy, and had shapes and curves without being in anyway fat. Blood still stained her face and her wimple.

'I fell.' She looked hard at Jimmy delivering a message. 'I fell and hurt myself. These two gentlemen were kind enough to stop and help me.'

Jimmy was impressed; the nun lied like a well-seasoned trooper.

The atmosphere changed immediately. Attitudes became friendlier and relaxed, hard men were suddenly gentle and considerate; handkerchiefs were offered. They wanted to rush her to hospital, she insisted the convent was fine, and they vied with each other to have her in their car.

Jimmy and Doc edged away from the group. Jimmy didn't know where to look first, at the guns or at Sean still hoaking about on the lorry.

Sean straightened with something in his hand. 'Mi...'

Doc's hand snaked out, caught his ankle and twisted. Sean yelled and hit the ground hard. He lay groaning. Jimmy bent over him, by the way helping. Sean had a gun in his hand; Jimmy scooped it into his belt.

Their questioner stepped out of the shadows and hauled Jimmy back, he was a big man. 'F....' His eyes caught that of the nun. 'Clear off Terence.'

Jimmy knew the voice now. It was that of the youth who had brought news of his father's death to Everton Secondary school.

He grabbed his licence. 'I'll see you again, Quinn.'

'You and which army?'

The nun came between them. 'I can't thank you enough.' Her

hands were pushing Jimmy to the safety of the lorry, even as her eyes pleaded for his continued silence.

Jimmy could see that they were lovely eyes, some light colour. He said, 'It was a bad fall missis. I'd see somebody about it.'

'I shall. And I'll write and thank you,' She indicated Mick who was helping Sean into a sitting position. Sean's head lolled and his mouth hung slack. 'This gentlemen will know your address.'

'Fuck I hope not,' said Jimmy.

★

The pubs were long closed when Jimmy got home. The lorry engine died before he could touch the ignition. 'Old bitch,' he said, and gave it an affectionate pat. He killed the lights and climbed slowly out of the cab.

'Still!' warned Doc, and disappeared into the night.

Jimmy stopped and listened, forcing himself to concentrate in spite of the weariness that clogged his mind. He realised that the lights in the house were on. He had switched them off as they left, he was sure of it, or he thought he was.

He saw Doc's shadow slide along the wall and look in the kitchen window. Doc straightened, laughed and walked openly to the back door.

Jimmy caught him up. 'What?'

'Goodnight,' said Doc. He said 'Goodnight,' again to somebody in the kitchen and went into the hall, closing the door gently behind him.

Jimmy found himself alone with their guest. They had been there long enough to put a match to the fire and use a bucket of coal to keep it burning.

'You must take me for a fool,' Eleanor said.

Jimmy was wide awake now and dying of embarrassment. 'You shouldn't be here.' He looked around, feeling there must be something he could do to make the place look better.

She stood up from the settee. 'When did you ever put work before me? Helping somebody? Yes. Doing a friend a good turn? No problem.' She flung herself at him. 'Don't get involved. Please, please. I'd die if anything happened to you.'

'I don't want to.' He said it into her hair.

'Then you won't do it again?' She looked up at him; there were tears in her eyes.

He felt he was betraying her. 'I can't.'

'Let the police do it, that's what they're paid for.'

He held her gently. 'They're only good for belting drunks on a Saturday night.' The fear he had felt earlier that evening, when surrounded by the IRA men, was still raw in his mind. 'This is war, the bottom line is our survival.'

She hugged into him again, and they held each other a long time. Eventually she scolded. 'You weren't so quick to come to my rescue the night the Troubles started.'

He was glad to get back to an old battle. 'I did, I cut round by your house. Your dad's car had gone so I knew the rats had left the sinking ship.'

'I'll give you rats.' She punched him, and then squealed as he up-ended her onto the settee. Her skirt went above her knees. It reminded him of the nun, he stopped there.

She struggled into a sitting position, her hand went to her dress and touched the hem, but left it high. He gulped. She said, 'In your dreams, how do you see me?'

'Like that. Lovely.'

She slipped a hand under her head. The movement flexed her breasts and did things to his stomach. 'Exactly like this?'

Having her there, not criticising the house, was like a dream in itself. He was wondering how he was going to get her home; the lorry wouldn't start again that night. 'Exactly like that.'

She knew he was lying from the blush burning his face. 'The truth.'

He licked his lips; he looked anywhere but at her. When he spoke he could hardly hear the words over the pounding of his heart. 'Well... without.' His hands fluttered at his buttons.

She unclipped the cameo and laid it safely on the ground, then took off her cardigan and undid the buttons of her blouse, the sleeves first. The skirt was loosened, the bra undone and everything slipped off and put aside.

'Well?' she said.

He licked his lips again. His hands started at his shirt buttons, the easiest to get at and the most innocent. He nearly fell getting his trousers and boots off.

He stood over her; afraid to do something that would spoil the moment. 'Is this your dream of me?'

'No,' she said.

Her arms were out for him before he could feel disappointment. He was grateful she seemed to know what to do.

She guided him into her. 'This,' she said. 'This.'

SUMMERLAND

Wendy found herself in a windowless cabin, lying on the ship's grey metal decking. She felt cold and shivery. Light, edged with red and green, glowed in that empty room yet had no particular source.

'Tony,' Wendy whispered. The thud of Tony's falling body still rang in her head. She was sure he was dead.

The walls started to move. At first she thought it her imagination, then the movement became pronounced as face after face pressed against the wall's thin membrane. She tried to pretend it wasn't happening, but the fear she felt behind the darkness of closed eyelids was worse so she watched, and saw those first faces withdrawn and new ones inserted, until a collage of death masks moulded in on each other, and a noise grew, of footsteps. Hurrying.

There was a faint tap on the door and a steward came in carrying a tray. He was middle aged and polite; he wore a white jacket with a high starched collar and smelled of aftershave, *Old Spice*. He had a pleasant smile. 'I thought a little tea, Miss.'

The faces against the wall and the footsteps had stopped with the knock at the door. Wendy stood up. Waiter service was something she wasn't used to but, nevertheless, it brought back a sense of normality. She wiped away the tears and brushed self-consciously at her skirt as the steward, Lithus he said his name was, put the tray on a table and hovered over it, ready to serve.

Where the table had come from she had no idea, nor the chair: a Victorian upright with a padded seat, its subdued tartan exactly matched the blue of her jumper. She thought she was going mad, had gone mad, and touched the chair half expecting it not to be there. It was, and she sat down because her knees were trembling. The steward gave her tea with a little milk and a twist of lemon, no sugar. It was as if he knew her foibles.

'There, Miss,' he said comfortingly, as she drank.

She needed a sip to clear her mouth and took a second to appear calm, afraid now that the tea might be drugged. 'Please let me go. I'll not tell anyone.' She knew her voice had come out high and nervous.

He looked up from tidying things on the tray. 'Miss?' And when she stood up he asked. 'Is Miss cold, a deck rug perhaps?' He offered her the one draped over his arm. Again, where it had come from....

'Please let me go. Please!'

'Unfortunately, Miss....'

He still smiled. She knew that when the smile slipped the danger would come, and she didn't know how she knew that. She wished she and Tony had gone anywhere that night but the abandoned cruise ship at the far jetty.

The steward made no attempt to stop her as she ran to the door and flung it open. She expected to find herself in a passageway and, hopefully, beside a flight of stairs leading up onto the deck. Instead, she found herself in a circular function room, the biggest function room she had ever seen, where a party was going on. A fancy-dress party. The smoke from burning candles drifted upwards to a ceiling lost in a brooding darkness far above.

A wall of sound stopped her dead in her tracks. String orchestras and quartets, balladeers and bawdy songs. The voices of elegant men bowing over their ladies hands. The whoops and hollers of country dancers. Laughter came from men and women seated at rough trestles. They used knives and teeth to tear steaming sides of meat into mouth-sized bites.

One woman went screaming around the trestles, pursued by three men. They caught her and tore the dress from her body. The diners sat and laughed as the now naked woman seized one of her pursuers and fell with him onto a trestle, where they copulated in a clatter of pewter plates and tumbling flagons.

The steward tut-tutted. 'These Skyclad people, I think they should send them on.'

Nearby, steel clashed on steel as two men engaged in a desperate sword fight. Wendy watched as one man slipped in under the guard of the other and struck for his opponent's heart. The injured man's mortal cry carried shrill and loud over the other noises. The sword slipped from his suddenly slack fingers and he fell. Onlookers exchanged gold coins in settlement of bets.

Wendy thought she was going to faint. The steward's supporting hand at her elbow stopped her knees from buckling. She couldn't believe it when the injured man got to his feet, and he and his adversary went off with their arms around each other's shoulders.

The steward murmured. 'They're always fighting. So typical of the Undersort.'

'Pardon?' asked Wendy, not understanding.

'The Undersort, Miss. There's the Milords, the Officers and their Servants. Then there's the Undersort.' His sweeping hand encompassed all those sitting at the trestles. 'It seems unfair that they should have an equal say in the order of things. Why if it hadn't been for his hidden pair of aces....'

Wendy looked around her and realised that the room was split into four distinct class groups, with only a minimum of contact between each. People watched back, and began to point her out and openly discuss her. She tried to retreat into the cabin and close the door against their curiosity. The steward's helping hand became a restraint. He still smiled.

He said, 'It's the Summer Solstice Ball, Miss. We always select a ship soon to join us.'

'We?' she asked, uncertainly.

'Of the Sargasso Fleet. If I may remind you, it's nearest port is in the Caribbean.'

Wendy looked at the people watching her and shuddered; some of the dancers had even stopped and were strolling in her direction. 'But what has that got to do with me?'

'As you can guess, Miss, virgins are at a premium. I think, having looked....'

'Yes, yes,' she said, and could hardly get the words out because her heartbeat clogged her throat.

The steward pointed. 'There, the three men in the corner, the ones playing cards.'

Wendy's eyes reluctantly followed his point. She saw three men at a green baize-covered table. One wore a wig and a salmon-pink coat with ruffles. He balanced an elbow on the table, the better to flash a rainbow of large-stoned rings. The second man was done up in a tight-fitting blue uniform of a Napoleonic sailor. The third man, obviously one of the Undersort, wore a rough leather jerkin of stained yellow, and a frayed shirt.

'They are playing for you, Miss. The winner….'

Wendy shook herself free of his grasp. 'When? I mean….'

The steward still smiled. She hated him for it. He said, 'It's a complicated game and the rules do not allow it to start until you board the ship. Regrettably, dawn comes earliest on the day of the summer solstice.'

'So?' She knew it stupid to feel hope.

'The party must end at dawn. All our pasts vaporise in the first light and we must go back to… Anyway, if the game runs to overtime or if you can delay the inevitable, then you win.'

'Win? It's hardly a game.'

She cringed, expecting to be punished for speaking so sharply, but just then the man in the wig and the pink coat gave a shout of victory, and the card-table expanded upwards and outwards to become a four-poster bed. Some force propelled Wendy forward.

The steward went with her and made the introductions. 'Monsieur le Marquis, may I present your companion for the evening, Miss Wendy Cowan.'

The marques smiled and bowed over Wendy's hand, the lace on the sleeve of his pink coat covered her fingertips. 'Mademoiselle, the pleasure is all mine.' He took a ruby ring from one of his fingers and slipped it onto hers. 'Please accept this small token of my esteem.'

Wendy turned and ran through the press of people. The men and women clapped and cheered as she clawed her way through them, but they were mostly the Undersort, and they made way for her pursuer who rapidly gained ground. Wendy found herself in a little gap in the crowd. Sharing the gap was the woman with the ardent suitors. This time they waited while she slowly and enticingly lifted her dress above her head. As she did so she pirouetted from one man to the other, choosing her partner for the next tryst.

The marquis came into the gap. Wendy grabbed the pirouetting woman and threw her at him. The woman held onto the marquis and said, 'Any man who wins the favour of the Skyclad is thrice blessed.' He bowed politely and she led him to the nearest trestle.

Now Wendy had another pursuer, the ship's officer. He was dark and swarthy, and wore his uniform with a careless panache. She ran. He walked briskly after her, saying, 'Madame, this is desertion. It is thy duty to serve thine elders lest cursed ye be.'

Wendy ran into Servants quarter of the room where they were doing Country and Western dancing. The men inclined their heads; the women gave her a curtsy-dip. They all politely made way for her. The Officer class were colder, wrinkling their noses in disgust at her bad manners, as she forced her way through them into the final section of the room where a quadrille was in progress.

The bewigged and corseted dancers closed in on her, trapping her in a huge circle, and danced around her is staid steps. The women flicked water at her from their pomanders. It tasted salty on her lips.

A giant black candle, set in a cowry shell, took up the centre of the circle, Wendy swung round it, going from side to side, looking for a way out. The dancers teased her. She ran for a little gap, but by the time she got there it had already closed and another one had opened at the far side. The officer stood outside the line of dancers waiting for her to emerge. A grandee in a lime-green silk coat, and with a black beauty spot on his cheekbone, spoke to him. The officer looked pleased at being recognised, and allowed himself to be led to the side and introduced around. The dancers moved off.

Everybody seemed to lose interest in Wendy. The swordsmen again clashed steel and the woman screamed round the trestles. Wendy sank to the floor. Her chest hurt from fear, and from running, and from the loss of Tony. She prayed for death and for her not to see it come. She wished, in all those oppressive grapplings with boyfriends at school, that just once it had been Tony and that she had given way.

They had met again the day her estate agency business acquired its first major customer. She floated up the street and into his arms. Her first words. 'You're still laughing at me.'

'Always,' he said.

In primary school she punched him for it, and in high school, her first day in tights, she hacked his shins. Eight years on he laughed with her, causing her whole being to glow. He felt the same, she could sense it behind his cheerful banter, and she was sorry she hadn't let him take her there and then in the street. Was even sorrier she had waited until now, the final night of his holiday, when there was no time to travel to a hotel where they wouldn't be known, and the only other place they could think of was the abandoned cruise ship.

Wendy looked up and saw the Undersort who had lost the card game, the man in the stained yellow jerkin, standing over her, smiling and showing his missing teeth. A sword point came out of the Undersort's stomach, he gasped and fell away to reveal the steward, sword in hand. Beside him one of the swordsmen protested that he had spoiled the fight and bad luck would come of it.

The steward picked Wendy up. 'I win after all,' he said, and carried her across the room and threw her onto the four-poster bed. A great crowd, a mixture of people from the four groups gathered round. The music had stopped, and there was only the beating of her heart and the crinkle of starch as the steward removed his jacket.

'Tony,' she whispered, and closed her eyes against a flash of white-fire from the steward's chest. She sensed a power trying to drag her along, and failing.

'I'm here,' said a voice. Tony's.

Wendy found herself in the ship's passenger lounge, lying on a settee. On one side, dock lights spilled in through the picture windows and, on the other, a first flare of sunlight speared the Eastern horizon. She put a hand to her head. Her hair was damp and smelled of sea-spray, other than that: no tears, no sore throat from crying. Her heart still pounded.

Tony joined her. 'Sorry I fell over my big feet.'

Wendy sat up and tried to make sense of the steward and the fancy dress party. She thought it couldn't be real. Yet, for a moment, a pier light became the moon-face of the steward, all angry and frustrated. Then it was gone.

Tony was saying how he had spoiled things, and that maybe they should wait until his next leave. Wendy put a finger to his lips, silencing him. She remembered the lost opportunities and the lovely man her tormentor at school had turned into.

She said, 'Now! You've no idea how dangerous it is being a virgin.'

WILFULLY AND DELIBERATELY

It was Geoffrey's first day back and he had forgotten a lot of things. He began to wonder if his daughter was right, that he should retire.

He said, 'Mr Anderson, how does your client plead?'

Anderson moved uneasily. He cleared his throat, the noise cut through Geoffrey's head like a buzz-saw. He coughed again. 'Not guilty, your honour.'

Geoffrey started and glared at Anderson who ducked down into his seat. Geoffrey drew breath and nodded to the Crown Solicitor, whose name eluded him. 'If you would present the evidence.' He settled himself to flick through the charge sheet while the policeman was sworn in: *absenting himself from confinement under an Electronic Tagging order; criminal damage; possession in suspicious circumstances.*

The Court Clerk had been at his most persuasive. 'A straight forward remand hearing, your Honour, in out and away.' No mention that he, the clerk, was anxious to get his own evening started, and now the young devil had pleaded not guilty. Geoffrey wondered if he should stand on his dignity as a County Court Judge, and insist that a magistrate be called back in to take over, and knew he couldn't.

The policeman, Constable Jackson, had nothing to say other than: 'Proceeding... apprehended... questioned...' Geoffrey thought him not a patch on his late father, Inspector Jackson. It only got interesting when Jackson talked about asking the accused to turn out his pockets.

'And?' prompted the Crown Solicitor.

'We found a piece of plastic on him.'

A piece of yellow plastic was fetched from the evidence-table and solemnly identified. Geoffrey handled it for form's sake. It was cut from a washing-up liquid bottle, slightly curved - he held it close to his nose - and still smelled faintly of lemon. That made him aware of something else in the courtroom, an aroma of smoke and stale drink, like a house the morning after a good party.

It came to him at last, Pollock. That was the Crown Solicitor's name; he had a drink problem and flat feet and a wife like a vixen. Geoffrey began to feel more confident and flexed his shoulders. *Who said he was past it?* He allowed himself a smile. 'Carry on, Mr Pollock.'

Pollock held up the piece of plastic for Jackson to see. 'You say you found this on the accused. Where precisely had he it...?'

Nasty, thought Geoffrey, *a jury would strain to finish that sentence: concealed; secreted; sequestered, and draw the obvious conclusion.*

Jackson said, 'In his hip pocket. It was all he had.'

'All he had what?' snapped Geoffrey.

'Pockets, your honour,' Jackson's face went sulky.

'One pocket, in a pair of trousers?'

Geoffrey looked at the defendant, a teenage boy in a filthy T-shirt. He shouldn't have, it made the boy nervous and he choked back a cough.

Geoffrey glared at the sound; the boy glowered back from under a cascade of untidy hair. *Surly little bugger*, thought Geoffrey, and felt better about the whole thing.

He turned his attention again to Pollock, the gnome and the window had to be presented for form's sake. They didn't have to bother, he knew what a window looked like from the inside: every mark, every dent in the frame, and the C and G intertwined; he had cut them into the glass himself. That was afterwards, after he fell in love with an old Edwardian house set deep in its own gardens, after he heard Cynthia choke down her fears in a cough.

They were still virgins; people were in those days, when they looked round their future home for the first time. They were in the master bedroom when Cynthia coughed back the tears. She didn't have to say anything: the faded brocades, the hand-stitched bedspread worn thin, the abandoned reading glasses all told of a lost future. They were silent for a long time. She cuddled into him and, as always, he soothed her tremors.

Geoffrey and Cynthia's fathers were partners in a firm of solicitors. The families were close friends. It was always assumed that Geoffrey and Cynthia would marry, their easy compliance, born of gentle natures, being taken for confirmation. The offer was now on the table, a junior partnership for Geoffrey and an interest free loan to purchase a gentleman's residence now in Executorship. The problem was that Geoffrey and Cynthia were friends, good friends. No more than that.

Cynthia looked at the bed again and coughed, nervous. Geoffrey took her hand. 'Lets try, and if we can't… you know… if it seems wrong. I'll call the whole thing off, say I changed my mind.'

'We both will,' she said.

Cynthia found brandy in the bedside cabinet. 'For medicinal purposes,' she said, charitably of the former owner. They used that, and there was only one chair in the bedroom so they sat on the floor facing each other, thigh against thigh. Cynthia did a lot of coughing as they settled. The first kiss was a tryout; they smiled when they tasted the brandy on each other's lips. It took a second snifter for them to get serious. Later on, much later, and still gloriously naked together, he used her engagement ring to cut their initials in the glass.

He sighed; it was the same window where he had stood for hours listening to the same cough while he watched the docken near his shrubbery grow and be cut down, and grow again. Anything else seemed so final somehow.

A fussy little man in a dark suit and club tie appeared, and was sworn in. He identified himself as *Mister* Edward Craig. Geoffrey reckoned Craig earned eighteen thousand a year, twenty at most, with children at the prep school, and mince and spuds for tea on high days and holidays.

Spuds? he thought. *Spuds?* He could smell them now, Susan's knocked up lunches and the onions crunchy under his teeth. *Spuds*, she called them, teasing him and his posh ways. And she laughed at his polite belch. 'You think that's bad, you should hear my flatulence,' he said. Cynthia wanted to know afterwards what they were laughing about. They told her and she was suitably shocked.

'My best friend,' Cynthia called Susan in the end, and she was. Others came visiting; Susan was there for the *gritty* bits. And she never flannelled him. The others said things like. 'Don't you worry, Geoffrey. Cynthia's strong, she'll fight this thing off yet.' Susan was blunter. 'Try to remember the good times together, it'll help afterwards.'

Susan was the materialisation of a vaguely heard name, committees and things: Parent Governor, Board representative. Not much to say, but worth listening to when she did. And a nice singer, he envied her that because his voice tended to wobble up and down the scales: Country and Western, folk, a bit of classical pop. She matched him song for song, and knew a few more besides: mostly the rebel sort, not done in *The Park*.

Craig thumbed in a tintack cough and identified the defendant, Peter Dunlop, as: 'a next-door neighbour, lower middle class. The adoptive father was a carpenter, what the real one was God alone knows.'

Pollock, the Crown Solicitor, asked, quickly. 'When did you last see the defendant?'

'About eleven fifteen last night. He was in the garden.'

'Did you not find that unusual?'

'No,' said Craig. 'He's always there.'

'At *night?*'

'He's supposed to be putting in a lawn for his mother, I've seen him at it until well after two.' Craig added, hastily, 'Not that I'm a nosy neighbour or anything, I like to keep my distance, but the wheelbarrow was squeaking and I asked young Dunlop to do something about it.'

'And the incident occurred at?'

'Near midnight. I was wakened by the sound of breaking glass.' Temper coloured Craig's face a rich ruby. 'The gnome hit the edge of the dressing table and knocked a piece of the veneer right off.' He glared at the Peter who looked back, the picture of innocence personified. 'That bedroom suite was top quality.'

Geoffrey thought, *Antique veneer, what a shame.*

Craig identified the remains of the gnome as his. 'It cost me a fortune, limited edition. *Harrods*, you know.' Geoffrey kept his eyes away from Peter; he couldn't bear the thought of another cough.

Anderson took over the questioning, and tried to position himself so that he could watch the judge, witness and Pollock all at once. Geoffrey thought it a good thing there wasn't a jury or Anderson would have had to turn himself inside out to keep an eye on them as well.

Anderson sidled closer to Craig. 'According to your statement, Mr Craig, you didn't actually see this act of...' He nearly said "vandalism," but caught himself in time.

Good point, Geoffrey thought, *pity about the fingerprints.*

Craig shifted uneasily. 'Well... no. There was the smoke you see.'

Stillness settled on the courtroom. Geoffrey found his eyes on the evidence-table, no smoke there: not a bottle, not a bag.

'Smoke?' asked Anderson.

'Yes, yes. Number 26, the Patterson's, nice people - Mr Patterson is something in the city - their house was on fire.'

Oh dear God, thought Geoffrey. *Arson! Whatever next? Murder at Number 28?* His head was pounding; it was all the coughs, all constrained. *There's Disprin in the wig-bag, and always something else, perhaps a Love-Heart.* That was Cynthia's job. And now there was no Cynthia.

He was bone-weary of this charade and wanted home. He leaned forward. 'Mr Craig, perhaps in your own words?'

Craig nodded to Geoffrey, a good sharp nod, the nod of one lover of justice to another. 'As I said, your Honour, the Patterson's house was on fire. If I hadn't called the fire-brigade when I did.' Craig's finger and thumb were almost touching. 'It was that close.'

'And you didn't go down to help?' asked Anderson.

'Not at that time. No.'

'I didn't think so,' murmured Geoffrey, and there was a momentary hiatus as he searched for words that would clarify the situation without overstepping the line. Words of Solomon. They wouldn't come, they never did, and Susan wouldn't believe him.

'Go on,' she said. 'An educated man like you, you'll always find some big word to suit the occasion.'

She was laughing at him. Every night she brought him a nightcap before she left. Often she would join him and they would talk. Sometimes of intimate things. Him, of golden handcuffs, and of a change in the rules allowing solicitors to become judges. She, of starting a business on a shoestring, and of rubber cheques travelling in both directions.

'Tut, tut,' he said.

'Tut, tut, yourself. I bet you know how to load a claim for Legal

Aid.'

He was at his most formal. 'I couldn't possibly comment.'

And they fell about laughing.

Cynthia never complained about missing the fun, she never complained about anything. At nights she preferred to choke back her cough and half-drown in the gathered sputum rather than risk waking him. So he'd learned to listen out for it, and for the restless movements when she was in pain.

He'd stand over her and pretend to be cross. 'Take the pills.'

'Can't like it.' She'd make a face like a petulant child.

Massage helped. He made himself busy collecting towels and oils while she draped herself over the edge of the bed; that way he didn't have to look at the scars. When she was ready he'd pound her back until the sputum came up. 'Cough,' he'd say. 'Cough.'

All she was fit to do was small and choked back. Especially near the end when she could only bear a gentle rub along the corrugations of her ribs. That was before the *Marie Cure* nurses, and the day nurse, and the night nurse, and the District Nurse with her morphine driver. After that, the only time they were alone together, truly alone, was around eight in the morning. He would sit on the edge of the bed, his thigh brushing hers – she found a touch too *heavy* for her wasted body – while he fed her Complan laced with brandy for taste. The only two things she could hold down.

Occasionally she asked, 'Is there anything left of the girl you married?'

He'd look at the worn face and the hair gone thin and grey. 'Yes,' he'd say with complete honesty. 'All of her.'

That final night Susan poured him a large brandy, 'Best thing I know to send you to bye-byes,' she said. She took a moderate one herself then joined him in a second. It was hard on her as well; her husband had died of cancer the previous year. The brandy worked, they both slept and came to curled into each other. Geoffrey could blame the drink, he could pretend that he was dopey with sleep, but when he tasted the brandy on her lips he knew that it was Susan sliding in under his body.

It was good to have breasts in his hands again and not be afraid of feeling compacted tissue. His bottom was bare and he was boiling in his jumper, Susan arched under him and the orgasm was sweet.

Then there was a cough from upstairs and reality intruded once again.

Cynthia died the next morning. The District Nurse was there for the final sigh. 'It happens that way,' she told Geoffrey. 'The heart gives out, best thing really. He couldn't cry, the relief that it was over, that she no longer suffered, was overwhelming.

The next time he and Susan met was at the graveside where she expressed formal condolences before going quietly, but firmly, out of his life.

Pollock coughed to attract Geoffrey's attention. Geoffrey started, he hadn't seen him approach the bench or Craig go. Pollock said, 'Your honour, in relation to the charge of non compliance with a Confinement Order.'

Geoffrey found his hands shaking as he accepted the document. He bent over it and began to read carefully. As he read, he became aware of footsteps in the courtroom, and of low voices interrupting each other. He ignored them all, only the printed words and their meaning were important.

Peter's whisper was loud, the word trembled with relief. 'Mum!'

Geoffrey's head jerked up, he wondered where Peter's mother had been all this time. Mrs Dunlop could never be thin, it wasn't in her nature, but she looked good. She was with Anderson and she was doing all the talking.

Another newcomer, a man, stood a polite step back from her, a fireman by his uniform and his hair was only starting to grey. Geoffrey envied him that; his was thin to the point of invisibility. *The fire, this man, the smell of smoke in the courtroom; there has to be a tie-in.* He took a chance and sneaked a quick glance at the boy. His surly air had gone; in its place was confident love. Pollock was trying not to appear interested.

Geoffrey nodded at the Defence Counsel 'At your convenience, Mr Anderson.'

The man took the stand and was sworn in. Watch Officer Ken Turnbull had been a fireman for fifteen years. He spoke the words of the oath in a voice that was as crisp and fresh as his uniform, and his nod to Peter was friendly. Peter blushed.

Anderson took a leaf out of Geoffrey's book. 'Mr Turnbull, perhaps in you own words?'

Not bad, thought Geoffrey, *he's a quick learner. We might make something of this young man yet.*

Turnbull didn't know what to do with his hat, he tucked it firmly under his arm, and he gave Peter another nod. Peter looked uncomfortable until his wandering eyes focused on Geoffrey then his face set into surly indifference. Geoffrey's hand itched.

Turnbull said, 'Your honour, I was Duty Officer with Blue Watch last night when a call came in at' - he checked in the book he carried – 'eleven fifty nine, a house fire at 26 Beaver Avenue. Respond time, 00.01 hours, arrived at scene 00.04 hours.

'We found heavy smoke and fire at the rear of the house. Also three people in the garden requiring medical assistance for smoke inhalation: the house owner, a Mr Herbert Patterson, his wife, Melanie, and a baby of about six months. Mr Patterson also had injuries consistent with falling downstairs.' He looked up. 'No smoke alarm, of course, some people never learn.' He sounded quite bitter.

'Mrs Patterson informed us that a second child, Louise, a little girl of four, was still in the house.' He consulted his notes again. 'At 00.06 hours, Fireman Little and myself donned our breathing apparatus and entered the premises. The smoke in the hall was almost solid and the heat was quite considerable. Flames were coming from under the kitchen door, and the architrave was on fire. I ordered a fire-hose on it and proceeded up the stairs.

'We found Peter Dunlop, the defendant, on the landing. He was suffering from smoke inhalation, but was still moving and had almost reached the top of the stairs.'

Peter was sitting with his head down. He wiped at his face and, for the first time, a hand showed over the lip of the dock, it was bandaged.

The little girl, Geoffrey thought, *what about the little girl?* He turned back to Turnbull and nodded for him to continue.

Turnbull said, 'With Dunlop, and being dragged along by him, was the missing girl. She was unconscious, but responded to treatment.'

'Thank God,' whispered Geoffrey, and asked. 'Can you tell the court how the defendant came to be in the burning house rescuing this four-year old child?'

Turnbull hesitated, uneasy at having to depart from script. 'What I know is hearsay but,' His voice quickened, 'it does form part of my official report.'

Geoffrey said, 'Very well, Mr Turnbull, you may continue.'

'Please,' said Anderson, trembling with righteousness at being bypassed.

'My apologies, Mr Anderson,' said Geoffrey, and thought him very like his father, a man to be watched. Though he had only seen Anderson Senior do one mean thing in his life.

With the coming of the Solicitors' Accounting Rules, a long established firm of solicitors in Kirktown, Moncrief & Co, was found to have a deficit in its Clients' Account. Even after everything had been sold that could be, there was still a massive shortfall in monies belonging to clients.

Come the white knight, Robbie Anderson, then a junior partner in a Dublin firm. For a mere fifty-five pounds paid to Moncrief, he took over responsibility for the shortfall and became the owner of one of the best practices in the county. Moncrief was saved from disgrace, humiliation and jail, but spent the rest of his life in comparative penury. Geoffrey often wondered about that fifty-five pounds.

Turnbull referred to his notes again. 'Mr Patterson reported that he first became aware of the fire, when he woke up to find young Dunlop standing over his bed shaking him. He has no idea how he got into the house.'

Geoffrey thought of the piece of plastic, and listened intently as Turnbull continued.

'Patterson went to fetch Louise, fell down the stairs instead and became unconscious. Mrs Patterson reports that Dunlop helped her and the baby out off the house then went back to rescue Louise. She did not drag her husband clear of the house, but remembers tripping over his body on the way out.' Turnbull closed his book. 'That is all I can tell you, your honour.'

'No questions,' said Pollock.

Again a hiatus settled on the court. Geoffrey waited; it had to come from the Crown Solicitor. *Hurry up before my head explodes*, he thought.

Pollock took his time about getting to his feet. 'Your Honour, the Prosecution accepts that the window and the gnome were broken with the intention of alerting Mr Craig to the fire at Number twenty six. However, two problems remain. The problem of the defendant's whereabouts from, say, midnight until this afternoon when he was apprehended by the good constable there.' He drew a reluctant breath. 'And his possession of an instrument designed and fashioned to force the locks of doors. To put it bluntly, your Honour, the tools of a burglar.'

Geoffrey looked at Anderson, Peter looked at his mother.

Mrs Dunlop said, 'Your Honour…'

Anderson quickly stopped her. They had a quick confab.

'I call Mrs Dunlop,' said Anderson, but by the time he said it she was already climbing into the box.

Geoffrey rested his head against the back of the chair and listened while she was sworn in.

'I, Susan Bernice Dunlop, do solemnly swear…'

Susan!

Red suited her, Geoffrey noted. Aware of his eyes on her she patted her hair into place. The movement stretched and straightened her body, and he remembered the softness of her breasts, no hollows where there used to be undulating curves. Maybe they sagged a little, she said they did, but that night they looked perfect.

He found himself smiling and straightened his lips. He knew he was down in weight and not eating properly since the operation, and that she'd noticed.

Geoffrey's head was turned towards Anderson, but his eyes remained locked on Susan. His mouth was dry. *Most unprofessional*, he told himself. *I should really adjourn this case and have someone else appointed.* Yet what could he say that wouldn't make him look stupid for letting things get that far?

Susan's colour was heightened, but she didn't cough or clear her throat. For that Geoffrey was grateful. She said, with eyes carefully averted from Geoffrey, 'My husband died of cancer two, almost three years ago. It was a long illness and I'm afraid, what with taking care of

Bill, that was my husband's name, and the business, Peter got overlooked.' The look she flashed to the dock was pure love. 'It was unfair on him because he has his own problems.'

'When you say "problems"?' That was Anderson.

'Peter was adopted.' Geoffrey and the court already knew that much. 'His natural parents were killed in a house fire when he was eighteen months old.' She stopped; there was a definite stir in court. Anderson palmed his hands upwards encouraging her to go on. Susan face became strained as old memories forced their way into her consciousness. 'Smoke inhalation, a bit like last night. Bill was too old for normal adoption.' Again the look of love flashed across to the dock, and was returned. 'We got Peter because he could never sleep before three in the morning, the time he was rescued; nobody else could cope with that.'

But you could, Geoffrey thought, *you could*.

'We coped.' That was Susan again, summarising years of disruption and exhaustion in the minimum of words. 'We thought things would improve as time went on, but it never did. No matter how tired Peter was, no matter how ill, he stayed wide awake until three in the morning.'

'Every night?' asked Anderson.

'Every night.'

'My God,' said Pollock.

'He kept falling asleep at school, and homeworks were a fight because he was exhausted in the evenings. When he got older, he started to slip out at night when I was asleep and got into all sorts of bother.' She made eye contact with Geoffrey for the first time, the steady stare she used to give him when discussing Cynthia's chances. 'This past few months Peter has been like an angel, going to the College during the day and staying home at night. When the weather's good he takes his restlessness out on the garden.'

Anderson hesitated before he asked the question. 'And the piece of plastic?'

Susan smiled, anywhere else she would have laughed. 'Sometimes I forget and lock-up on my way to bed. Peter's gardening trousers are... well I'm ashamed he's in them today. They're definitely for the bin this time.' That was a challenge across to her son. He gave her a wicked grin and shook his head. 'They've only got one pocket and that's full of holes so he can't carry a key.'

Anderson breathed a sigh of relief. 'So the piece of plastic has quite an innocent purpose?'

'Yes. I'm afraid it's a little trick he picked up when he was running the streets. He never used it,' she added, hastily and, for the first time, was an anxious mother instead of a calm, confident businesswoman.

Anderson went for broke. 'Can you account for your son's movements between, say, midnight last night and this afternoon when he was apprehended by the police on his way home?'

Susan acted surprised. 'Didn't he tell you?'

Everybody in the room looked at Constable Jackson, who ignored them and stared blindly ahead. *Oh dear*, thought Geoffrey, *oh dear.*

There was a hard set to Susan's mouth. 'Give a Dunlop a bad name....' The look she directed at Jackson was decidedly unfriendly. 'Somebody had to know, you're not all stupid.'

Geoffrey didn't dare intervene. Anderson said, hastily, 'Please.' Susan stood silent, her head high and neck stretched. *Ruffled*, was the word that popped into Geoffrey's mind.

Finally she relented. 'When I heard the sirens I went looking for Peter. I knew he'd have gone to help, and I didn't want him caught out of the house if the probation officer rang. Somebody told me he had been hurt.' Her voice wavered at that point then strengthened again. 'It was a cold night and they were using a house away from the smoke to treat the casualties in. He was asleep.' She stopped then repeated it. 'He was asleep. On the floor, in the middle of all the noise and fuss. It was hardly midnight.' She stopped again, and used her thumb to wipe moisture from beneath her eye. 'They wanted to take him to hospital, but the neighbour, bless her, said it was all right for him to stay. He was on his way home when.... Anyway....'

Geoffrey asked, 'And that was the first time ever?'

'Yes.'

There was just him and her, everyone else was forgotten. 'Do you think...?'

'I daren't even begin to hope.' She looked towards Peter and the magic moment had gone.

'Thank you, Mrs Dunlop,' said Geoffrey, in his most formal voice, and she was dismissed.

He waited until she had taken a seat, until he could draw breath. His brain had developed a euphoric lightness that transcended the pain. When he got home he would take a medicinal brandy and sleep himself better. In the meantime he didn't dare look at either the boy or Susan until he came to a decision.

He knew that technically Peter was wrong for leaving his garden to rescue the Pattersons. *Forget that.* But he was wrong for not going straight home afterwards. The piece of plastic was a serious contravention of his probation terms, and there was a dozen ways the boy could have raised the alarm. Lobbing the gnome through Craig's window had more than a hint of sweet revenge to it.

He said, 'Peter Arthur Dunlop, you are ordered by this court to be taken direct from here...' He stopped and looked into Susan's eyes and saw the sparkle of tears. *Serves you right*, he thought, *you let him take chances he shouldn't have.* He continued. 'From here to the Area General Hospital, where you will remain for such length of time as the doctors, in their good wisdom, decide. At which point...' His eyes took in every man in the court, not one of them was breathing. 'At which point, you will return to and be confined to your home in strict conformation with the original terms of your probation.'

Susan was smiling in relief. She touched a hand to chest, lips and forehead and gave the smallest of bows.

SCORING

'That's another party survived,' I said.

Moya smiled. 'Thanks for the lift, Steve.' She touched my arm as she spoke, but her hand was away before I could trap it.

I followed her out of the car and put my arm round her waist. Her body stiffened then relaxed against mine.

I muzzled her ear. 'How about a coffee? I'm tired and I've a fair bit to go.'

I was too. Office parties when you're not drinking can be boring, but that way I got to take the girls home. Usually they travelled in twosomes, that night Moya had manoeuvred to be the last one dropped off. I was surprised at that, I didn't think she was back on the market and, even if she was, I had never seen myself as her type. She got the door opened and stepped into the hall with the little hop she always gave before blasting the ball at the net. We were in the mixed football team at work. Moya was a forward and I played goals.

The hall had subdued lighting coming from under the open stairs, nice and romantic. I took her in my arms; she let me caress her back, but blocked my lips with her fingers. 'You, sitting room, me kitchen,' she said, and broke contact, and went into the kitchen with her head tilted up - the way she did when she was thinking or was annoyed about something.

I decided not to push things, but went on into the sitting room where a photograph of her late husband grinned at me from the mantelpiece. The heat of the room was pleasant. I obeyed her shout to switch on the television. I couldn't find the controller so did it manually. Then I made myself comfortable on the settee and let my eyes close. Too many late nights were catching up with me.

The tap-tap of Moya's red pointed shoes on the wooden floor roused me. She came in carrying two mugs of steaming coffee and put them on the coffee table. The mugs were exactly the same, that made me feel uneasy, then I noticed the spoons: hers was from Ibiza, mine from

Crete. She sat down beside me. I relaxed and put a hand on her knee.

'Keep it for stirring the coffee,' she said.

I laughed. 'I have to at least try.'

'You've tried. Now leave it at that.'

She meant it too. The thinking in the kitchen had gone against me. It added to the spice of the night – and to the night of the previous office party when I had taken her on the settee. Not that she remembered anything about it. The next day at work she appeared hung-over and puzzled about something. I played it cool and friendly and the moment passed.

Moya watched me as we drank. There was a tenseness about her; her eyes kept flicking between the television and me. I was thankful that she was drinking slowly, the more coffee in the mug the less chance of her getting the bitter taste of the pill.

If I did it.

I had this feeling that maybe twice with the same person was dangerous, but her blouse spilled open every time she leaned forward, revealing breasts that were firm and generous. To generous to resist. This time I would undress her completely.

'Coffee's awfully dry on its own,' I said.

She buried her face in her mug, unwilling to move.

'Even a digestive?'

'If you want.'

She got up and went to the door. I could have killed her; she was carrying her mug. She stopped in the doorway and looked at me, frowning. I smiled and turned back to the television, apparently more interested in the adverts than in her body. I wondered if I dared do a bit of slapping around, no bruising mind. She came back and left her mug on the coffee table. As soon as I heard tins rattling in the kitchen, I shook a pill out of my inner pocket and popped it into her coffee. I stirred it well in with Ibiza, and gave my head a good shake at the same time to keep the tiredness at bay.

She came back with a plate of things. I took a handful of biscuits, but left my coffee on the table, hoping she wouldn't think to drink her own until the tablet was well dissolved. My mouth was dry and my heart fluttered in anticipation.

Moya sat well away from me, her head tilted up, listening to the television. A new programme came on, a feminist programme about the increasing problem of date rape. Moya sat frozen, watching me out of the corner of her eye, as the women on the television gave their horror stories. With their haggard looks and lack of pride in their general appearance, they should have been grateful in the first place.

I couldn't believe my luck. If Moya did claim rape, I could say her intoxicated mind was muddling herself and me with the stories she had heard. I don't think she had taken much that night, no more than a couple of vodkas and Coke, but who could remember for sure?

Moya lost her stiffness; she might even have edged towards me. 'Would you mind turning the television up a bit?'

I put a hand on her thigh as I stood up; she went all spiky. I backed off smiling. 'Just testing the defences.'

When I got back she was drinking her coffee, a good gulp I noticed. I checked that I had Crete and drank as well.

We sat listening to a debate on date rape. Mostly the experts talked rubbish. A pill slipped into their drink gave women what they were gasping for without them first having to pretend unwillingness.

I was so into the programme, I almost missed Moya subsiding into her corner. Her head went down and she gave a little snore. I put a hand up her skirt, her breathing changed, but she never moved. I was struggling with my own tiredness as I pulled my trousers down and her skirt up.

She slapped my face. I fell off the settee, and bumped my head and felt dazed. I saw a woman's legs in front of me: black tights, black shoes, and knew that couldn't be right.

I looked up and found myself staring at another woman, a police sergeant with the build of a bull elephant. She spoke into her radio. 'Barlow here, send a car. We've got the bastard bang-to-rights.' Then she pointed at the television and said, 'Don't bother trying to deny it, we'd a camera on you the whole time.'

I staggered to my feet, my legs felt rubbery.

Moya was on her feet as well, straightening her clothes. 'Are you all right?' asked the policewoman.

Moya replied, 'Fine, Vera, I swapped mugs and spoons.' Then she took a little hop forward and drove for my groin with her pointy red shoe.

Mercifully the tablet kicked in.